"As a business executive, leader, coach, and mentor, I am always looking for educational opportunities that help me, and the people that I work with explore new opportunities that will expand the collective knowledge base or moreover create that inspirational feeling that helps you drive to new levels or reinforce existing best practices. Bill Artzberger has created that opportunity in his new book, *Fostering Innovation.* Bill masterfully and specifically takes the reader on a journey of failure and success in the lives of several fictional leaders and employees, all the while underscoring the basic tenets of innovation, Lean theory, and principles.

The journey starts with John, a company CEO who has created a 'crest' for his company that embodies the use of innovation to drive differentiation, creative thinking, and ultimately growth for his company. Throughout the book, John is both challenged and supported to learn that fostering innovation and creating an environment that embraces the principles of Lean require more than putting words in a mission statement or on a plaque. Bill's use of 'real-life' storytelling along with bold print that accentuates the key learning elements allows both the experienced and novice leader to grasp the concepts of creating a work environment that supports creativity, risk-taking, and ultimately success.

Fostering Innovation is a must read for any leader or manager who is focused on the use of Lean to help drive innovation and transform the culture of their business. The development of the characters and the flow of their story allow the reader to quickly engage and self-reflect. Bill's tactical use and examples of Lean theory and process throughout the book force the reader to clearly see the impact of the principle and the consequence of making the right choice. The questions at the end of each chapter allow for a guided education and contemplation of the key learning points in the book. The glossary at the end of the book provides a solid compendium of terms and definitions for a leader at any level."

– **Barton P. Buxton**, EdD,
President and CEO – McLaren Health Management Group

"*Fostering Innovation* is a practical approach in exploring leadership as well as group and team development. Bill has successfully connected the core principles of Lean principles to innovation. I found the book to be an effective tool that applies academic concepts to organizational situations. The case study approach makes the read enjoyable, reinforcing innovation as a core competency. Readers have the opportunity to examine communication, teamwork, leadership, diversity, creativity, reward/recognition, and culture. I highly recommend this book as means of self-examining oneself as well as organizations for sustaining innovation. Steve Jobs once said, 'Innovation is the only way to win.' As a retiree from AT&T and currently the director of the Pawley Lean Institute at Oakland University, also serving as adjunct faculty, I find this book relevant for personal growth and as a supplement to academic coursework in the fields of Lean, leadership, and group/team development."

– **Dennis L. Wade**,
Director – Oakland University Pawley Lean Institute

"Like Goldratt's classic, *The Goal*, Artzberger's book refreshingly teaches and reads like a novel instead of a textbook. The ideas and concepts underlying *Fostering Innovation* are presented as a real story of real people struggling to figure out their careers and path to success. Lean and continuous improvement and leadership, all parts of innovation, are not presented as chapters. Rather, they show up as solutions to the problems at hand. The result is a better read, and a better way to learn and retain."

– **Randy Hinz,**
General Manager – Automotive Manufacturing

Fostering Innovation

While innovation can be defined in many ways, the author sees it as a process. It is not the sudden eureka moment in the middle of the night, nor is it a clear and linear path toward a final destination. Instead, it involves a strong sense of creativity and curiosity. An innovative mind has a natural inclination toward out-of-the-box thinking. It involves a willingness to try something new, without fear or judgment, to develop something no one else has ever articulated. While the mindset comes naturally, it requires fuel to keep it running. Innovators are voracious readers and researchers. They feed their mindset all of the fuel it needs to stay informed and relevant in their field.

Many of the same things can be said for the Lean mindset. Lean management doesn't happen overnight, and it is very rarely a clear and linear path to true Lean thinking. Some might consider Lean a subset of innovative thinking, while others see it in reverse. Regardless of the relationship's directionality, one thing is certain: you cannot have one without the other.

This book follows John Riley, the CEO of a medium-sized valve company just outside of Pittsburgh, Pennsylvania, who will stop at nothing to create an innovative work environment. Through the ups and downs of his journey, he learns a number of Lean and innovative skills, strategies, and mindsets to help him build the business he's always envisioned for himself.

Throughout the book, you see examples of both strong and poor innovative leadership skills demonstrated by each of the main characters. The key messages are ones that help leaders build and access a mindset insistent on continuous improvement. Leadership techniques and abilities that bolster creative thought and problem-solving are the most successful throughout this book.

To be truly innovative, you can never stop driving the learning process. For this to happen, leaders need to recognize when there is a need for a change or improvement. This is the beauty of the marriage between Lean and innovation: they both require continuous learning and growth. The desire to improve is only one piece of this equation, however. The other is the willingness to act. Without both of these factors, true innovation will always be out of reach.

Fostering Innovation

How to Develop Innovation as a Core Competency and Connect the Principles of Lean in Your Organization

A novel by

Bill Artzberger

Edited by

Brittney Leigh Schlechter

Illustrated by

Patrick Harrington

A PRODUCTIVITY PRESS BOOK

First published 2023
by Routledge
605 Third Avenue, New York, NY 10158

and by Routledge
4 Park Square, Milton Park, Abingdon, Oxon, OX14 4RN

Routledge is an imprint of the Taylor & Francis Group, an informa business

ISBN: 978-1-032-33136-2 (hbk)
ISBN: 978-1-032-33135-5 (pbk)
ISBN: 978-1-003-31835-4 (ebk)

DOI: 10.4324/9781003318354

Typeset in Minion
by SPi Technologies India Pvt Ltd (Straive)

Contents

About the Author

Bill Artzberger, PMP, LSSBB, is the managing partner at the Lean Learning Center. He specializes in Lean manufacturing, innovation, Lean healthcare, leadership, and project management.

He has over 35 years of experience in real-world senior management, including CIO, VP, president, and CEO. He has worked with thousands of individuals from the boardroom to the shop floor, in virtually all sectors of industry. Bill holds two patents and brings extensive project management, Lean manufacturing, and information technology experience to the Lean Learning Center. He has published dozens of times and is a co-author of the top-selling Lean book *Driving Operational Excellence* and author of *Powering the Lean Enterprise*. He is also an associate professor at Oakland University and a member of the OU Pawley Center advisory board.

He has successfully implemented Lean improvement projects in dozens of companies throughout the world. Bill is an expert in the usage of Lean techniques such as value stream mapping, error proofing, quick changeover, Kaizens, Kanban systems, 5S, and visual pull systems. He has taught hundreds of Lean process classes and is an expert at driving large-scale operational change programs.

Bill earned his MBA from Wayne State University in Detroit, Michigan, with a major in management and a minor in marketing. He received a BA from Walsh College in Troy, Michigan, with a major in information systems and a minor in finance. He also holds an AS from Oakland Community College in Farmington Hills, Michigan, with a major in systems analysis. Bill is ITIL certified, a certified Lean Six Sigma Black Belt, and a PMI certified Project Management Professional (PMP).

Acknowledgments

Thanks to all the talented people at the Lean Learning Center and our fabulous clients for their help and endless ideas.

Thanks to Patrick Harrington and my family for the endless support and suggestions (and patience).

Special thanks to Brittney Leigh Schlechter, whose insight and tireless effort made this book possible!

Introduction

FIGURE 0.1
Bright new ideas bulb

While *innovation* can be defined in many ways, most people see it as a process. It is not the sudden eureka moment in the middle of the night, nor is it a clear and linear path toward a final destination. Instead, it involves a strong sense of creativity and curiosity. An innovative mind has a natural inclination toward out-of-the-box thinking. It involves a willingness to try something new, without fear or judgment, to develop something no one else has ever articulated. While the mindset comes naturally, it requires fuel to keep it running. Innovators are voracious readers and researchers. They feed their mindset all of the fuel it needs to stay informed and relevant in their field.

Not long ago, the assumption was that innovators were born. They came into the world with a natural ability to see things differently, work through problems more strategically, and navigate adversity with more ease than their less innovative counterparts. Researchers weren't convinced that this was true, and a new field of study began. Since then, research has shown that innovators are developed, not born. They utilize a number of skills, strategies, and mindsets to see the world with curiosity and creativity.

This book follows John Riley, the CEO of a medium-sized valve company just outside of Pittsburgh, Pennsylvania, who will stop at nothing to create an innovative work environment for his organization. Through the ups and downs of his journey, he learns a number of these skills, strategies, and mindsets that help him build the business he's always envisioned for himself.

In Chapter 1, we meet Alexander Filmore, a recent college graduate with impressive skills and talents. He is driven and ambitious. As a company that desires to be at the cutting edge of its industry, John feels that Alexander will make a perfect fit.

That is, until Alexander starts using his creativity in ways that don't align with John's vision, John has made a name for himself implementing the same ambition and drive that Alexander exuded during his first few days with the company. However, as John's experience and expertise grew over the years, he began to clasp tighter and tighter to the notion of doing things "his" way. Among other things, John's headstrong ways and toxic culture create a barrier to Alexander's growth and ability to exercise his innovative talents.

In a last-ditch effort to find his footing at TrinoTech, Alexander takes advantage of a professional development opportunity that addresses the small steps toward building innovation in a company. Through this program, Alexander is exposed to the concepts of psychological safety and minimizing bureaucracy. The longer the program continues, however, Alex learns what he fears the most; that John is not ready to loosen his reigns and let his company shine.

In Chapter 2, we see a slightly different side of John. He feels completely blindsided by Alexander's swift exit and is struggling to address the concerns Alexander mentioned in his resignation letter. Andrea Alastor, John's mentor and AlaStar's CEO, attempts to coach John around several learning and development strategies. She explains the importance of developing culture and the value that experimentation can bring to a company, even if risks are involved. She also encourages John to investigate coaching and mentoring opportunities that his company might consider implementing to bolster the retention of his top-performing employees.

As the conversation builds, John is forced to face himself in the metaphorical mirror Andrea is holding, reflecting John's ability to lead an innovative company. While she certainly doesn't doubt his capabilities, she does question his mindset about innovation and future-oriented thinking. She requests he answer one simple question: what do you want your company culture to look like? John leaves the conversation feeling both confused and frustrated but determined to make things right at TrinoTech.

In Chapter 3, we meet Michelle and Mary Lynn. Michelle is a new employee with TrinoTech, and Mary Lynn works for AlaStar, the company owned by Andrea. As the two friends share a few beverages, Michelle informs Mary Lynn that she is still unhappy at work. John doesn't seem to value Michelle's creativity and opportunistic mindset. Instead, he wants his people to do things his way with very little consideration for others' ideas. Mary Lynn sympathizes with her friend and tries to offer some advice she's learned since working for AlaStar.

She explains the value of social capital and how Andrea demonstrates her appreciation for those who work for her. Mary Lynn reminds Michelle that innovators are proactive and persistent. They like to challenge the status quo and push for a better future. Unfortunately, these qualities tend to fly in the face of leaders who aren't ready to step out of the spotlight. Still, she encourages Michelle to stay strong and encourages her to attend a Lean conference the following week.

In Chapter 4, we see a glimmer of hope in TrinoTech's ways. John haphazardly creates a learning and development budget and encourages his employees to use it for professional development purposes. Though there are still a number of things John needs to think through, it finally feels like a step in the right direction. Without hesitation, Michelle asks to use the money to attend a Lean conference, and her supervisor is happy to oblige.

The Lean conference is both insightful and informative. The two women, along with Alexander and Andrea, uncover the truth behind Lean and innovation's relationship, learning that Lean is essential to innovative progress. Through the four Lean rules, the participants dive deeper into how Lean enables innovation. First, they learn that eliminating chaos by adding structure is key to early problem detection. This clears the mind for more creative thinking about solving these problems. The second rule encourages companies to draw clear lines between their suppliers and their customers. This serves to further eliminate chaos and frees folks up for more creative and innovative thinking. The third rule builds on the second by requiring simplification for each flow path. The last rule, however, is the most challenging: improving through experimentation. Without the successful implementation of the first three rules, however, the fourth rule will likely fail.

Though Michelle learned a few new techniques and strategies for innovation at the conference, they go largely ignored at TrinoTech in Chapter 5. John continues to struggle with innovation, trying unsuccessfully to implement and control a variety of Andrea's suggestions. Andrea realizes that John is burning his people out by only pulling in high-level staff to execute the new ideas. In addition, she sees that he is pushing his team to identify and develop one large idea instead of focusing on developing some of the better, but smaller ideas. She reminds him that innovators recognize everyone's voices at their company, not just those at the top. They are also good at clearly communicating expectations and parameters around new projects. She recognizes that John's biggest hurdle is that he is afraid to let go of the major project ownership for fear of losing his credibility. She encourages him to implement an after action review (AAR) process to analyze how he's performing in the eyes of those who work for him.

Chapter 6 brings us back to Michelle and Mary Lynn at Panera, where Michelle discloses her increased concerns about John's behavior. John is trying to make changes to advance the company but is struggling to recognize his employees' concerns. Mary Lynn recommends a culture audit so that John can gain a deeper understanding of his company's needs. Mary Lynn can see that John is trying to do right by implementing small changes into the company culture, but that he's not treating innovation as if it's a core competency. She mentions Lean's core principle of developing a learning organization and explains the mindset that goes into building this kind of Lean culture. She encourages Michelle to find a way to talk to John about building motivation by leveraging people's intrinsic desires and company values for success.

In Chapter 7, we begin to see a real shift in John and his mindset around innovation. He realizes that he's not handling change well, which is likely because he is taking on too many things at once. Marcus, the owner of HydroTherm, leads a seminar that challenges John to question whether he's truly ready to embrace the changes he is trying to make at his company. After listening to Marcus explain the Innovation Adoption Curve, he finally realizes that he needs to take the transition seriously or risk falling behind the competition.

Marcus continues the seminar by discussing the benefits of brainstorming and idea implementation. He strikes a chord with John when he gets to his discussion on Learning Systems: that the only way to sustain the growth and changes he would like to create is to establish a strong learning system. As Marcus continues his seminar, John finally connects with all of Andrea's advice. He realizes that it's time for him to focus on taking the next big steps with his company.

In Chapter 8, we see a completely invigorated John. He creates a personal mission and vision statement and starts executing them by recruiting employees into positions based on their skills and talents. He asks Michelle to head up a Knowledge Management Committee that will be responsible for developing a company-wide knowledge management system. Michelle eagerly accepts and works tirelessly with her team to research and develop the right kind of technology for the company. Ultimately, she ends up proposing a risky option for the company, though she strongly believes that it will be the best option for John to consider. While John feels initially torn, he sees the value in Michelle's decision and accepts her proposal.

Alexander, Mary Lynn, and Michelle have become close friends since their first Lean conference together. They meet at a local pub in Chapter 9 to discuss a big news for Michelle: she got a promotion to Director of Continuous Improvement! As Michelle tells her story, Mary Lynn recognizes Andrea's influence. She explains to the others that Andrea is the gold standard for servant leadership in the Lean industry and highlights how she used her power and influence to help John make changes at TrinoTech through the Five E's of Innovation. Without Andrea, Mary Lynn posits, TrinoTech would not have grown into the strong company it is today.

In Chapter 10, we travel ten years into the future and get a sneak peek at what Michelle and John have been developing. Michelle, now the Vice President of Learning and Development at TrinoTech, has developed an extensive internal learning plan. She offers online learning opportunities and encourages employee knowledge shares, mentorship opportunities, and think tanks. Now, TrinoTech is thriving in a way John never imagined possible for him or the company.

Throughout the book, you will see examples of both strong and poor innovative leadership skills demonstrated by each of the main characters. The key messages are ones that help leaders build and access a mindset insistent on continuous improvement. Leadership techniques and abilities that bolster creative thought and problem-solving are the most successful throughout this book.

To be truly innovative, you can never stop driving the learning process. For this to happen, leaders need to recognize when there is a need for a change or improvement. This is the beauty of the marriage between Lean and innovation: they both require continuous change and improvement. The desire to improve is only one piece of this equation, however. The

FIGURE 0.2
Mindset candle

other is the willingness to act. Without both of these factors, true innovation will always be out of reach.

These mindsets don't necessarily come hardwired in one's brain at birth. Instead, people who seek to be innovative can cultivate these mindsets and develop the skills and abilities they need to create innovative ideas and environments over time.

Part I

What Is Innovation and How Does It Work?

1

Defining Creativity and Innovation

FIGURE 1.1
We have always done it this way

DOI: 10.4324/9781003318354-2

Alexander Filmore barely noticed the gently falling snow as he walked into the front lobby of TrinoTech. A gold emblazoned crest curved around a scripted TT and was inscribed with the words: *The Creative Mind and Innovative Spirit Move Us Forward.*

Yes, Alexander thought, *this is where my future begins.*

It had only been a few weeks since Alexander's interview, but he was confident that TrinoTech would be the place he built his career. John Riley, the founder and CEO of the company, valued creativity and innovation above all else.

"At TrinoTech," John had said during Alexander's interview,

> we believe that **creativity** is a discipline. **It's the ability to think up new and unique ideas.** Using one's imagination to **think outside the box and come up with something not yet thought of** is admirable. How would you describe your level of creativity?

"That's a great question," Alexander had responded.

> At my internship, I was consistently praised for my **forward-thinking and ability to develop fresh ideas.** During my exit interview, my supervisor emphasized that my creative input helped move the company forward. I believe creativity is one of my top strengths.

"That's excellent," John replied.

> Of course, as you can tell by our crest, it doesn't stop with creativity. It's one thing to be able to think up new ideas; it's another to **develop and implement** them. There are many ways to define **innovation**, but we see it as **a process.** Many companies are innovative for several different reasons. Disney is innovative for its storytelling abilities; Toyota is innovative for its Lean thinking. Here, we define **innovation as the ability to identify unmet needs and introduce progressive change** to our products. We value the work it requires to bring an idea to fruition, and we do not shy away from it. Given this definition of innovation, how do you see your personality fitting in with our organization?

Alexander could feel his nerves vibrating. Finally, a company that truly understood what it meant to be innovative. So many companies he had interviewed with had mentioned innovation and creativity only briefly during the interview process. Here was someone who truly seemed to care about these two aspects of the technology field.

"To be perfectly honest, sir, I think this company would feel like home to me," Alexander responded.

Two days later, he received the call that he was the number one choice for the design engineer position at TrinoTech. He almost shouted at the HR woman when she offered him the role. His dream was finally coming true.

He walked up to the front desk and introduced himself to the receptionist. If he was going to be working here for a long time, he wanted to learn everyone's name. She smiled kindly at him and helped him find his office. Well, it was more like a glorified cubicle, but Alexander didn't care.

Then, she walked with Alexander to the other desks in the engineering wing of the office and introduced him to the busy men and women who made up the rest of his team. He was excited to see the bustle of his colleagues, collaborating on projects and drawing concepts using the software on their computers. Everyone seemed engaged in their work and eager to come up with the next new idea.

Alexander was ready to join them.

MYTHS OF INNOVATION

The first few weeks at Alexander's new job started just like any other job. He worked hard to come up with new ideas, contributed his thoughts in staff meetings, and managed to make a few new friends in the office. He was no stranger to late nights, spending much of the after-hours time coming up with unique ideas to propose to John at the next staff meeting.

During the first few meetings, John barely afforded Alexander the time to speak unless it was to update the group on an existing project. In the last meeting, however, Alexander knew he needed to speak up when John glossed over a design issue the engineering team was having with one of the valves.

"Sir," Alexander began, "I believe we are looking at this wrong."

The look on John's face told Alexander everything he needed to know about people speaking out of turn during a staff meeting. A few seconds passed as John decided how he would handle Alexander's unsolicited advice.

"What makes you say that, Alex?" John finally asked. He set his pen down on his tablet and leaned back in his chair, crossing his arms over his chest for added effect.

Alexander took a deep breath and reminded himself that he had been exclusively working with the product they were discussing. He had been trying for weeks to implement some of the suggestions that had come out of these meetings with no success. He was better equipped than anyone in the room to answer the question.

> Well, sir, I believe we are trying to solve the wrong problem. We keep changing the material the valve is made of to hold more pressure, but I think we are actually looking at a design flaw.

The room fell silent. John himself had been a big part of the valve design and had personally reviewed all of the drawings.

"I think we need to take a step back and **make sure we are defining the right problem first**," Alexander added, breaking the silence.

John nodded his head slowly and leaned forward, lacing his fingers together on the folio pad in front of him.

> Thank you, Alex, for your interesting assessment. At our company, we admire an innovative spirit, one that seeks to solve problems, *not* create more.
>
> Yes, sir, I understand, but I think this is an opportunity for us to be more innovative with the design of our valve. We just need to rethink how the design could accommodate extra pressure, not the materials. I've already come.

John held his hand up to stop Alexander from speaking.

> Alexander, again, I thank you, but **I am not interested in problems. I am interested in solutions**. We have discussed several solutions to this problem in our meetings, none of which require a redesign. I admire your ambition, **but I have been at this a lot longer than you have**. I reviewed this design many times. I know it will work with one of the solutions already proposed. Your job is to make that happen. Do you understand?

"Yes, sir," Alexander responded, feeling dejected.

The rest of the meeting went on without another interruption from Alexander or anyone else for that matter. While he did have a few other ideas he wanted to discuss, he chose to remain silent. All he could think about was the way John had shot down his ideas so readily, without any evidence to back up either of their claims.

"Do you think he doesn't trust me yet?" Alexander asked Deliah, his cubicle mate, after the meeting ended.

"I don't think it's about trust per se," Deliah responded. "He just thinks that his solutions are always the right solutions since he's been doing this for so long."

"Isn't that just as bad as **the eureka myth**?" Alexander asked.

"The eureka myth?" Deliah asked.

FIGURE 1.2
The Eureka myth

> Yea, **the idea that there needs to be some kind of ah-ha moment for something to be innovative. Most innovative ideas are developed over a long period, not just thought of in one minute**. It's one of the myths of innovation, just like thinking you know better simply because you are the boss.

"I guess I hadn't really thought of it that way," Deliah shrugged.

"And **wanting to focus on solutions, not problems**," Alexander continued, "**is another myth. Defining problems is often *more* important than the solution itself.**"

"I don't disagree with you, Alex," Deliah said, "but he has his way of doing things."

"I understand that. I guess I was just wondering when we talk about new things. You know, 'the creative mind and innovative spirit move us forward'," Alexander reasoned.

"I think you'll find that the creative mind and innovative spirit only move us so far, and are only created in one person's image," Deliah said.

FIGURE 1.3
Overnight success

"What do you mean?" Alexander asked.

> Well, I mean that John talks a lot about innovation, but he also relies on his "**innovative method**". He often talks about his **"innovative process", which allows him to be creative and forward-thinking**. He even teaches a quarterly training on his "innovative process" as a way to help us think in more innovative ways.

Deliah explained. "Which basically means, he wants all of us to think more like him."

"But that's just another myth," Alexander replied.

> During my internship, I went to a conference on innovation in tech fields. One of the workshops was on innovation myths; things that people often think innovation needs to be but never is. **There is no such thing as one correct "innovative process". Creativity and innovation are messy processes. It's not really something that can be boxed up into a method and taught as a step-by-step process to follow**.

"Well, don't tell John that," Deliah laughed.

> He clearly feels the opposite. He follows his method without fail. I think the number one thing I've heard since I've started is "**at my last company, we…**" or "**we've always done it this way**". It's not the most forward-thinking, I know, but he is very dedicated to it. He struggles to **let go of issues from the past**. Last year, one of our more junior engineers, Mick, developed an impressive new Bluetooth software for one of our valves. John shot it down right away without even looking at the specs. He said TrinoTech tried Bluetooth before, and it didn't work. Mick left a few months after that.

"Didn't HydroTherm come out with a Bluetooth valve earlier this year?" Alexander asked.

"Sure did," Deliah responded.

John was furious at first, but when he realized that it was similar to Mick's design, he completely disregarded it as a competitive product. "This is very similar to the one we tried a few years ago. It won't last a month on the market before it starts failing," he said. "**The best predictor of future outcomes is past outcomes.**" He ultimately decided he didn't want to upgrade the valve at all.Alex smiled widely at Deliah.

"Let me guess," she said. "Another myth?"

"You got it," Alexander replied. "**Progress is never a straight line. It's often muddled and unpredictable**. An idea that didn't work before can very well work in the future if it's properly vetted, especially if the problem with its first iteration is adequately defined."

"So, all of these myths overlap too?" Deliah asked.

"Well, not necessarily," Alexander answered,

> but you can see how one false belief can breed others. I guess I'm just surprised. **I know that another myth of innovation is that people get excited about new ideas when the reality is that most people naturally resist change**. I didn't expect it here, though. John seemed like someone who really valued creativity and innovation. I mean, I know I just started, but I also know that I have a ton of ideas that have probably been tried before in one way or another. That doesn't mean I can't make them work now.

"I just think that John is also naturally skeptical of good ideas," Deliah replied.

> Although this one I know is a myth because John talks about it all the time. **He knows that good ideas exist everywhere; he's just not great at encouraging them**. I'm sure you just surprised him by being more outspoken today. People don't typically offer too many ideas in his meetings without his prompting.

"Well, one thing I'm not scared of is speaking up," Alexander pondered,

> but I don't want it to get me in trouble either. It makes me nervous to see so many of these myths popping up at this company. I was excited to be working somewhere that valued creativity and innovation. I haven't seen a whole lot of that so far.

Deliah wasn't sure what to say. She enjoyed working with Alexander and didn't want to see him leave the company before he got more out of it. Alexander had already improved so many aspects of their department. She knew losing him would be a huge hit to their office.

"It's possible you could come up with something he does like," Deliah said. "Just because I haven't seen it yet, doesn't mean it can't happen."

"I hope you're right," Alexander said.

After heading home that night, Alexander considered his conversation with Deliah. He didn't want to be naïve, but he also hadn't gotten any one-on-one time with John yet. Perhaps, he thought, a meeting to talk through some of his ideas would help push things in the right direction.

Alexander fired up his computer and began typing an email. It was time for him to have his first meeting with John.

BARRIERS TO INNOVATION

John delayed meeting with Alexander for two weeks. It's not that he didn't have the time; he just wasn't sure he wanted to entertain a one-on-one meeting with Alexander after his performance in the staff meeting two weeks prior. In those two weeks, however, Alexander continued to offer new ideas during staff meetings. A part of John loved that Alexander was not deterred by John's remarks on the valve, but another part of John felt frustrated. TrinoTech's valves have consistently done very well in the market. Why did Alexander feel the need to challenge that? Why couldn't they stay the course, especially on things that were already going well?

When the day came for his meeting with John, Alexander arrived five minutes early. The delay in getting a meeting with John had allowed him to prepare adequately. He created a thoroughly researched presentation for why he strongly believed a design change for the valve was necessary as well as a list of questions about company culture. Still, as prepared as he was, he felt confident he would be met with some skepticism.

"Good afternoon, Mr. Riley, thank you for taking the time to meet with me." Alexander extended his hand.

"You can call me John, Alex." John leaned forward out of his seat lazily and immediately sat back down. "I understand you would like to review a few schematics of the valve drawing. Is that correct?"

"Yes." Alexander removed a small folder from his backpack and pulled out the original drawing. He pushed it across the table. "As you are aware, this is the original drawing. This," he pointed to the casing around the valve in question,

> is what we have been trying to fix. At this point, we have tried three different materials for the valve casing, and the closest we've come to making it work is a more pliable material, like rubber. The first test with the rubber material almost always passes, but each test after, the numbers get worse.

Alexander handed John a document with all of the test results for the valve over the past few weeks. John looked down at the paper with a furrowed brow, but he was only pretending to listen. He bided his time, waiting for the appropriate moment to thank Alexander for his time and send him back to working on the valve based on its original design.

Alex again reached into the stack of papers in front of him and pulled out a second drawing. He passed it across the desk to John and leaned in closer.

> This is the new drawing I came up with. I have been working on the valve for weeks, and none of the material changes are helping. Even if it passes the first test, it won't pass the second. I don't think we should force a product into the market that has a high fail potential. So, I stepped back and realized that part of the valve design is causing pressure issues. If we make the valve itself a sixteenth of an inch wider, I think we can use a rubber casing safely.

John stared down at the new drawing in front of him. By all accounts, Alexander was right. How could he not have seen this himself? It was such a simple solution to the problem, but it was also uncommon to change the valve width. He'd only ever seen it done once before.

"I don't know, Alex," John heard himself say. "I still think my design will work."

He wanted to like Alexander. The reason he hired Alexander was that Alexander reminded him so much of himself, but he certainly **didn't want anyone who would step on his toes.** This was his original design, after all.

Hearing his own **jealousy** revolve in his thoughts, he decided on a different approach.

"**We have a certain way of doing things around here,**" John started again.

> During your probationary period, the expectation is for you to learn from experienced design engineers; people who have the experience and know what they are doing. Many of our engineers, myself included, have been doing this for years. We know what will work and what won't. I appreciate your drive, but **this is the way it's always been done**.

Alexander pondered his options. On the one hand, he respected John and his company. After all, John had built this company from scratch only a few years ago. He must know a lot more than Alexander ever could. On the other hand, Alexander knew that John's attitude would prevent the company from moving forward at a more rapid rate.

"I do respect the processes and procedures that you have in place here," Alex began.

> I also think that your design would, under any other circumstances, be an incredibly strong product. In this case, however, we haven't been able to make it work. I think that we need to be looking at alternate options. I know the design I came up with will work, but if you prefer to have another senior engineer redesign the valve instead, I am fine with that. Either way, the design as it stands won't work moving forward.

John looked down at the two drawings in front of him. Both drawings were essentially the same, with one miniscule difference in valve diameter. Doing quick math in his head, John realized that Alex was right. While his design would work nine out of ten times, it wouldn't work for the kind of pressure release this particular client required. Still, he wasn't convinced that Alex's design was the best option either. They had tried a valve with a larger diameter at one of his past companies, and it didn't work the way it was intended. There must be some way to make the original design work.

"Listen, Alex, I know that you are young and eager to prove yourself." John leaned back in his oversized chair and stroked his dark beard.

> Your creativity is one of the reasons we hired you, and I appreciate you taking the time to come up with a potential solution. Still, I believe that sometimes experience trumps creativity. We tried a valve similar to the one you are proposing at one of my past companies. It didn't work out. I am not going to stand by and watch history repeat itself on my dime.

Alexander considered the **cultural precedent** John set. It was becoming more evident by the second that TrinoTech's culture was one of veiled creativity and innovation. Sure, they were a groundbreaking company a few years ago. Some of their valves had been some of the most innovative in more than a decade. Yet, as the meeting drug on, Alexander could see John's fear. **TrinoTech got ahead by trying new things, but as soon as those things started becoming profitable, there was no longer a need for creativity and innovation. The trend quickly became: *stay the course*.**

"Speaking of money," John continued,

> **we don't have the money or resources for a redesign this late in the process.** If we want to fulfill our promise to our client and deliver by the end of the first quarter, we need to move forward with what we've got. If this was two years ago, I might have kicked this to the R&D team. However, they have since been dissolved to make room for positions like yours. Positions, I might add, that are expected to get things done and stay on deadline. **We don't have the capacity to kick this around**.

Silence fell between the two men. It felt thick and sticky as Alexander decided on his next move.

"Sir," Alexander said a bit more quietly,

> I looked over the budget. I am confident that I can make the new valve design work without any additional resources, money, or time. I am very concerned that the design as-is will give us problems over the long run. I'm worried about potential leaks. I know I can make the new design work. I just need your approval.

John took another deep breath. Alexander was beginning to wear him down, and he didn't like how that felt. Alexander should have taken his word as gospel at the meeting and gone back to work. Talk about a waste of resources. They had spent over an hour discussing one valve design when Alexander already had everything he needed to make it work. John glanced back over the drawings once more. While he was certain Alex's design would produce a viable product, he felt there was no way for him to back down now. He was the owner of the company. His word was final.

"If the valve passed the first test, then there is a way to ensure that it passes the second," John said with a harsh finality to his tone. "We hire engineers like you to make things like this work. Now, please go and make it work."

Alexander slowly got up from his chair. He gathered his documents back into his folder and, with a brief nod at John, walked out of the office. Not once did John ask for more information on the new valve design. Not once did John even consider the potential of a different design. John made it clear to Alexander that it was **his way or the highway. No matter how good they are, innovative ideas won't fly at the company without coming from John himself**.

"How did it go?" Deliah asked when Alexander returned from John's office. "You were in there for a while."

"Not good," Alexander said. "I think I'm going to head home for the day."

Alexander packed up his desk and walked out of his office. He reflected on his initial interview when John sounded so open-minded and excited for new ideas. His meeting with John showed a very different side of John, which was much more **closed-minded and self-righteous**. How would they ever grow with a closed-minded thought leader?

SMALL STEPS IN INNOVATION

In the weeks since his meeting with John, Alexander grew more and more unmotivated. He liked the kind of work he was doing, but he felt unheard and unappreciated for his talent. Alexander started worrying that he was falling into the same trap as some of his working friends: apathy due to poor leadership. He knew that motivation and job satisfaction rates plummet when leaders don't take time to recognize or listen to their employees, and he didn't want to fall into those statistics. He knew he needed to do something to boost his motivation and get his head back in the game.

At first, Alexander started looking for other positions. He kept replaying John's words over in his mind and decided that working at a company that doesn't value his innovative spirit was not a place he wanted to be. However, Alexander had only been with TrinoTech for a little over six months. He couldn't leave so soon. He wanted to give it more of a chance than that.

After a few days of scrolling through lists of suggestions for folks suffering from low motivation, he decided to take an online course. One of his top character strengths was a love of learning. He knew it would boost his motivation if he could learn something new and find a way to incorporate those new skills at work.

He decided on an online course called *Seven Small Steps to an Innovative Work Environment*. The course was self-directed, which allowed him to digest the content at his own pace.

"When I first started teaching about small steps in innovation, I received a lot of puzzled glances." A recording of Doctor Chuck Alder,[1] a leader in innovative theory and practice, said at the beginning of the course.

> You see, many people think that innovation is a grand and remarkable process towards the next biggest invention. Misconceptions around innovation have pervaded conversations around this topic for decades, and many of these misconceptions actually slow down the flow of innovation. **One of the most pervasive myths is that of the lone inventor, the person who walks into a lab with a brilliant idea and pops out the product completely on their own**. Thomas Edison, Albert Einstein, and Alexander Graham Bell are all extremely good examples of the "lone inventor" stereotype. Of course, we know that each of these men had an incredibly talented and brilliant team of people behind them, helping make their innovations a reality. Sadly, we don't hear much about these people. They rarely get credit in school textbooks, nor do their names typically appear in museums. **It is the negligent withholding of such information that encourages people to assume that innovation is a gigantic stroke of luck and brilliance carried out by a singular entity**.

Alexander related to everything Dr. Alder said. At work, it was clear that John wanted all of the credit and accolades for himself. His meetings were often riddled with anecdotes from his time inventing one of the first TrinoTech valves, and he loved to remind folks of the effort he put in to build the company from the ground up. If anyone bought into the idea of the "lone inventor" misconception, it was John.

> Truly, it takes a team to bring an idea – any idea – to successful completion. But where does that innovative spirit and mindset come from? This is where leadership comes in. While anyone can follow the steps in the course, it is imperative that leadership be fully invested in sustaining an innovative culture.

Alexander let out a deep breath. *Is this even worth my time? John will never be invested in innovative strategies.* Alexander scrolled down to the first module: Mission and Vision. *TrinoTech does have a mission statement that revolves around innovation. At the very least, this can't hurt my future.*

MISSION AND VISION

"The first step in sustaining an innovative culture is **understanding why innovation matters to your business,**" Dr. Alder began. "What will innovation allow you do to more of? What is the purpose of your company's innovative mindset?" Leaders should ask themselves questions about purpose, direction, and goals regularly to continuously grow their business. Asking how innovation fits into that purpose, direction, goal, etc., is a foolproof way to generate clear and concise messaging for the rest of your company.

> Once leaders can answer and assess some of these tougher questions, they should consider their mission statements. Some leaders might consider including ideals around innovation in their mission statement. A mission statement needs to capture the essence of the business. If innovation is a major part of how a company functions, leaders should consider making that a key message within a mission or vision statement.
>
> Finally, once determinations are made about the mission or vision statements for a given business, leadership should consider how that messaging is received. Ideally, it should be written somewhere public to serve as a daily reminder of what each person is working towards.
>
> Take a few moments now to **consider why innovation matters to you**? Perhaps a good place to start is asking yourself why you decided to take this class in the first place. Consider some of the questions I asked earlier and jot them down in a journal or notebook. Then, work through each section of the module until you have a working mission statement for yourself or your business. Finally, consider where you might display your mission. If you lead a company, you might consider making it a banner on your company website. If you are just getting started or going through this course as an individual, you might consider taping it to your computer screen or bathroom mirror. It is important to put it somewhere that you will see it frequently.

"Once you've completed the exercise, you can move onto the next module: Psychological Safety," Dr. Alder finished.

Although Alexander was not the lead of the company, he worked his way through the module as an individual. While he couldn't make a mission statement for his company, he could be sure that a mission statement for himself was accurate and thoughtful. Who knows, perhaps he would consider starting his own business someday.

PSYCHOLOGICAL SAFETY

"Step two focuses on a concept called Psychological Safety," Dr. Adler began when Alexander was finished with his first module. This concept has been thoroughly researched by many scholars, most notably by Amy Edmondson. Essentially, **psychological safety is the degree to which a person feels comfortable (safe) taking interpersonal risks with another person**. While this includes close, personal relationships, it has been most widely studied in the workforce.

> Psychological Safety is an essential piece of workplace innovation. It encourages open dialogue and communication and bolsters the way members of a team collaborate. It means that all team members feel comfortable about offering ideas or opinions. In fact, this is encouraged and even expected. There will be no repercussions from speaking out. Nobody gets fired or shunned in any way for offering an opinion. They also feel comfortable about asking for help. Open and honest communication is always encouraged. No one will experience a CLM (career limiting move) for engaging. Everyone feels safe about speaking their mind, even if the idea does not turn out to be a good one.
>
> The key here is trust. Employees want to feel a high level of trust, not only with one another but also with leadership. This means that **leadership should also maintain a reasonable amount of consistent communication across the company**. Did the company hit a milestone early? Send out an email. Are tariffs going to cause a price increase on a product in development? Send out an email. Yes, it's important to send updates about the negatives as well. The more employees know, the more comfortable they feel with their leaders.

Alexander shook his head as Dr. Alder explained more about sharing negatives. While John did a great job of sending out correspondence highlighting the positives, he would certainly never send out an email with a problem. Earlier this month, his lack of communication caused a big problem with the sales team. The China tariffs caused a hike in price for a certain kind of steel from overseas. The owner of the China-based company sent John an email to let him know that prices were on the rise. However, John didn't share that information with sales, who continued to sell the valve at the same cost. TrinoTech essentially broke even on the part for a few weeks until our lead saleswoman caught the issue and adjusted the price.

There are many things leaders can do to foster psychological safety with their teams. One of the most significant pieces of advice I give to leaders is to **remain engaged**. As leaders in this fast-paced economy, it can be easy to get distracted by phones or computers. Putting those things away during a conversation with an employee can go a long way to helping that employee feel heard and understood.

Demonstrating a sense of understanding is another thing leaders can do to bolster psychological safety. This sounds simpler than it is in practice. Obvious things like repeating or summarizing what someone said, or using validating statements are typically easy practices. What's harder is avoiding the urge to place blame on someone when something goes wrong. Leaders often get themselves into a state of auto-pilot and can sometimes forget to stay focused on solutions.

Another big piece of psychological safety is inclusion. Companies who value innovation also value collaboration and constructive criticism from across the company. **Leaders should always seek to include employees in both interpersonal settings and when making decisions.** Leaders should always be approachable, express gratitude, and be available in interpersonal settings. Employees should always feel like their opinions are valuable and heard by leadership.

For this exercise, I encourage you to make yourself a chart. In the left column, write down the words engagement, understanding, and inclusion, each in their own box. In the right column, grade yourself on in each of these categories. Are you always engaged and not distracted? Do you seek to understand the folks you are working with? To what level do you include others in both interpersonal and decision-making settings? Once you grade yourself, I would encourage you to make a list of strategies you want to use to get stronger in each area.

In the next module, we will discuss how Lean strategies can help eliminate an abundance of bureaucracy.

As Alexander assessed himself in each of the categories, he was surprised to learn that he had graded himself low in the inclusion category. He concluded that he was likely too eager to prove himself and didn't do a great job of asking his colleagues for help. He made a note to put his phone in his desk when talking to anyone at work and ask his coworkers more questions about the upcoming projects.

MINIMIZE BUREAUCRACY

The third step involves limiting bureaucracy. Now, I know what many folks are thinking. Shouldn't leaders be the ones making all of the decisions? Not necessarily. This is where Lean strategies can come in handy. **One of the central purposes of Lean is to eliminate waste: anything that doesn't add**

value to the customer or product. One of the things companies have found in implementing Lean methodology is that decision-making bureaucracy has become a big bottleneck for the company.

If you aren't currently considering Lean, I highly recommend that you do. Some of the most innovative companies in the world practice Lean strategies with astonishing results. If you are implementing Lean strategies, I recommend looking at policies and procedures that involve approvals. Do too many people need to sign off on a single decision? Who are the key stakeholders?

This is an incredible opportunity to practice psychological safety. Call a meeting with all of the people working on a project or send a survey out to your organization to assess where they perceive delays in decision-making are occurring.

In the meantime, take a moment to think through an upcoming project. How long, from idea to implementation, will it take? How many people will that idea pass through before it is funded? Once it's funded, how many people need to sign off on the final implementation? If you are noticing an excessive amount of time for each stage of the project, perhaps it is time to reassess.

The next step will review employee autonomy and creativity.

Alexander paused the video to think through an upcoming project. As far as he could tell, at least four people needed to approve it before it could be added to the next team meeting agenda. Then, assuming the team reached that part of the agenda during the meeting, the whole team would need to discuss it. Once discussed, John would allow himself two to three weeks before making a final determination, which he would then send to the finance office for final approval. Even with final financial approval, any tweaks or design changes would then have to go through the same process. It could take months to get a project off the ground. Alexander took a few notes that he planned to discuss with John during his next one-on-one meeting.

AUTONOMY AND CREATIVITY

"Leaders will often overlook just how important a sense of ownership is to an employee," Dr. Alder said as he began step four. **Instilling a sense of ownership encourages employees to look for creative ways to add value**

to the company. There are many different ways to provide a sense of ownership to employees, but what it boils down to is helping employees see how their efforts make a difference.

One way to instill a sense of ownership is through stocks. **Employee-owned businesses tend to grow faster, pay better, and have higher job satisfaction rates than traditionally-owned businesses.** Giving employees a way to invest in the future of the company naturally makes them more likely to seek out ways to help the company grow.

Don't have the ability to give employees stock ownership? No problem. **Another way to instill a sense of ownership is to give employees the time and space to come up with their own unique ideas.** For example, Google allows its employees to devote 20 percent of their time to their own creative projects. Truly innovative companies will find ways to give their employees time to be creative.

Now, while stock options and time for creativity are both great options to provide autonomy and ownership to your employees, the most powerful way to accomplish this goal is to allow employees to **own their own creative process**. I'm sure you're wondering how that's different than giving employees time for their own creative projects. **Give people the creative autonomy to change and adjust their own working process, to not be stuck with what was handed to them, and to truly own the outcome of their work.** The biggest kiss of death in any organization is "we've always done it this way". Forcing employees to perform a task in a way that does not engage them is a quick way to lose good people. Let them own the process that helps them arrive at the best possible product.

Do you offer stock options to your employees? What do they look like? If it is possible to offer stock sharing to your employees, I would strongly consider it as a way to drive employee buy-in. Even if it's not possible now, perhaps it is time to think about a plan for making it a possibility in the future.

In the meantime, consider how you can add creative time to your employees' schedule. Perhaps 20 percent of their time is too much given the size of your company. Could you afford to give them 10 percent? 5? In what ways can you encourage your employees to act in more innovative ways? You may also want to consider how much time you give yourself for this kind of work. Leaders who give themselves time to innovate are more likely to instill those same beliefs in their workforce.

Finally, consider how often you are using the phrase "it's always been done this way" or "I prefer if you do it this way". Are you truly letting your employees have ownership of their workflow? Consider giving your employees more autonomy and space to complete their projects in a way that is both engaging and productive.

In the next module, we will cover recognition and rewards.

Based on this step alone, TrinoTech was far from being "truly innovative." In the three months Alexander had been with the company, he had been working around the clock on assigned projects. It was in his spare time and late nights that he worked on his own creative projects. Even so, he was never granted the space to propose or discuss those projects. If anything, he felt that his drive for creativity and innovation was seen as more of a burden than a company asset. While the exercise for this module wasn't completely relatable to Alexander, he did consider how much of his free time he spent in creative thought and made a note to provide himself more creative space every day.

RECOGNITION AND REWARDS

"How does your company recognize and reward innovative thoughts and ideas?" Dr. Alder asked his invisible audience.

Here we go, Alexander scoffed. *This should be interesting.*

> One of the biggest mistakes I see companies make is rewarding one team or one individual for the successful completion of a new idea. Let's say, for example, that a large tech company just came out with a very popular tech device, perhaps a new cell phone or music player. It would be easy to go ahead and reward the design team responsible for the product. But what about the marketing team who made it relatable, or the sales team who made it profitable? What about the administrative folks who fielded phone calls and filed all of the paperwork? It's easy to see how **rewarding only the design team might demoralize the other teams who had a hand in making the product successful (even if it wasn't their idea). This can establish a culture of competition.**
>
> As we mentioned earlier, one of the first steps in an innovative work environment is collaboration. It should seek to recognize innovative ideas no matter their magnitude. As we mentioned in the last step, it's important that every employee, no matter their role, understand that they add value to the company. **An innovative culture is one in which the rewards and recognition process is one that continues to foster collaboration and teamwork.**
>
> One of the biggest questions I get in response to this step is, What about the people who are truly excelling? This is where it helps to have rewards that don't revolve around financial incentives. **When a company notices that an**

employee is excelling, the best thing it can do for both the employee and the company is to cultivate the employee's skill set. Consider paying for them to attend a conference or training in their respective area of expertise. If the budget is too tight, consider pairing them with a more experienced person with similar traits at the company so that they can continue to learn more. The key here is to **show the employee that the company is just as invested in them as they are in the company**.

Sit down and write a list of the last three people/teams you recognized for their work. Ask yourself the following questions:

- **Did I recognize *everyone* who played a role in that success?**
- **Do these instances range in magnitude?**
- **Did I offer professional development as a reward?**

If you answered "no" to any of those questions, I would encourage you to sit down with the rest of your leadership team to discuss a more streamlined way to recognize exemplary teams and employees. If you don't already have a professional development budget for your company, I highly recommend setting aside some money this year to reinforce your incentive programs.

Another failure for TrinoTech, Alexander thought. Although he hadn't been at the company very long, he had asked around about how teams are recognized for their work. For the most part, he was met with confused gazes or blank stares. As Alexander worked his way through the course, TrinoTech got further away from what an ideal innovative work environment should be.

"Our next step is all about risk and failure," Dr. Alder said.

Great, Alexander groaned. This is only going to get worse.

RISK AND FAILURE

Step six is one of the easiest steps to explain, the hardest to implement, and is closely linked to almost every previous step in the process. In the world of Lean, we call this "Improve through Experimentation" and is one of four Lean rules. **Essentially, to have an innovative work environment, risk and failure need to be encouraged and accepted**. For many leaders, however, risk and failure are scary things,

Dr. Alder explained.

"However, the truth of the matter is simple: you don't get innovative ideas without taking any risks. In taking risks, there will inevitably be a few failures along the way." **The key here is to see those failures as an opportunity for learning and growth, not as an excuse to stop taking risks.** That's what continuous learning is all about. If a risky decision leads to failure, consider asking the following questions:

- **What did we learn from this?**
- **What can we do differently to prevent this from happening in the future?**
- **How will we grow from this?**
- **What aspects of this risk should we continue to work on moving forward?**

> With big risks often come big rewards, but these larger risks rarely work perfectly the first time around. The company willing to put in the time and resources to keep pushing an idea down the line is the company that will reap the bigger rewards. **Continuous learning is key to a Lean and innovative work environment.**
>
> Risk-taking is a value instilled from the top down. In step two, we discussed the concept of psychological safety. If employees don't feel safe, they won't take any risks. Therefore, **the leader must demonstrate an appreciation for risk-taking and an understanding that a few failures in the process are inevitable.**
>
> Unfortunately, leaders see experimentation in a silo instead of seeing it as an opportunity for improvement. By allowing your folks to experiment with processes and procedures, you are **assessing the current reality within those processes. Understanding those realities will help ensure that past mistakes are not repeated and help drive those processes toward an ideal state.**

The easiest way to assess this for yourself is to think about the last time you took a risk at your company. Ask yourself some of the following questions:

- **In the grand scheme of things, how big was the risk?**
- **Who was involved in taking the risk?**
- **What goals and structure did you set around the risk?**
- **Did that risk help drive you toward an ideal state?**
- **What kind of flexibility did you communicate to your team?**

If the risk was relatively small, with no goals or structure, and poor expectations about your own flexibility, you likely communicated a lack of trust in your team. If risk-taking is something that scares you, consider starting with smaller risks as a way to practice communication and flexibility. As you grow more comfortable with risk and failure, allow yourself, your team, and your company to take bigger risks. You never know where they might lead you.

This one was easy for Alexander to assess from TrinoTech's position. Since he started working there, they had not taken any risks. They simply continued to work with the same or similar designs they have worked with before, and, as Alexander so painfully realized, the risk was something that was certainly not encouraged.

In our final step, we will revisit how Lean strategies can strengthen innovation.

With a heavy sigh, Alexander hit play on the last step.

LEAN STRATEGIES

Lean methods contain these three components:

- **Waste elimination**
- **Added value to end-users (customers/clients)**
- **Continuous improvement**

Each of these pieces can have dramatic impacts on innovative work environments.

We touched on waste elimination a little bit in step three, where we discussed eliminating bureaucracy. In Lean, however, waste elimination goes even deeper than that. For example, **Lean calls for simplicity in processes.** It requires the removal of unnecessary steps that may result in confusion or errors. If something is no longer working, consider asking yourself why it's not working.

As mentioned in the last section, experimentation is crucial. **Another rule of Lean is to structure every activity to make processes repeatable.** In the case that something isn't working correctly, consider setting up an experiment. Assess the effects both upstream and downstream. If it works, great! The experiment was successful, and the process has been improved. If it was not successful, that's also great! The experiment then eliminated the potential for repeating mistakes, and you can set up another experiment. Either way, you have fostered an environment of innovation by eliminating chaos from the workflow. Doing this continuously adds value to the product and company.

We've also touched on the idea of added value in steps five and six. When employees have a reason to invest in the company, they will ultimately provide more value back to the customer/client. A deep understanding of mission and a sense of psychological safety creates added value by eliminating tribal knowledge and enabling access to the collective knowledge of the entire group. This is where innovation happens.

Finally, the most interesting aspect of Lean for innovation is the idea of continuous improvement. You might argue that striving for continuous improvement is also striving for innovation. **Lean requires companies to be in a constant state of assessment, which helps foster and develop a learning organization**. Auditing systems, tracking employee feedback, and eliminating wasteful practices are all opportunities for innovation. These strategies enable employees to fully engage in their work, making them more likely to make improvement suggestions that contribute to the growth of the organization.

As I said before, if you aren't currently practicing Lean at your company, I highly recommend doing so. **Innovation is sure to follow the implementation of Lean strategies**. If you are practicing Lean, consider reassessing some of the areas we mentioned at the beginning of the module. Ask yourself:

- **When is the last time I audited my processes and systems?**
- **When is the last time I checked in on employee feedback?**
- **How often do I hold space for employee feedback and ideas?**
- **How am I actively reinforcing the continuous improvement mindset and learning environment at my company?**

As you can see, all of these steps rely equally on one another for an innovative work environment to thrive. You can't truly have an innovative work environment without each of the steps mentioned in this course. Enforcing some steps while ignoring others is like trying to craft a puzzle without all of the pieces. I would encourage you to do a deep dive into your company culture to assess where you stand on the innovation spectrum. Innovation is a continuous improvement process. It requires constant vigilance and a desire for your company to grow and develop.

"I thank you for joining me for this course."

With that, Dr. Alder signed off.

Alexander's head was spinning. He had hoped that this course would give him perspective and maybe even a few ideas he could take back to propose to John. After watching each step, though, he realized that the problem of innovation at TrinoTech was much deeper than he had realized. Innovation is a top-down effort, one that requires buy-in and energy from the leaders of the company. While Alexander had only worked with John for a few months, he hadn't gotten any indication that John was interested in changing his ways.

That night, instead of writing up a report on the course and creating a presentation to pitch to John, Alexander scoured the internet for other job openings. This time, he looked through each company's reviews and only applied to companies whose employees mentioned innovation and creativity as a key aspect of their culture.

Within two weeks, Alexander had a job offer from HydroTherm, a valve company in a different industry with a four-and-a-half-star employee rating. Their CEO was a mission-driven man who often lead innovation and creativity training and workshops all over the country. It was the fresh start Alexander needed.

QUESTIONS FOR DISCUSSION AND REFLECTION

1. How do you combat the myths of innovation at your company?
2. What innovation myths were you most surprised to read about?
3. Are you or your company guilty of perpetuating any of the myths?
4. How do you combat innovation barriers?
5. What innovation barriers are you most guilty of falling into?
7. What small steps in developing an innovative work environment can you or your company take?

8. How do you know if your employees trust you?
9. What do you do to enable an environment of psychological safety?
10. When is the last time you assessed the success of your innovative work environment? Where could you improve?

NOTE

1 Doctor Chuck Adler is a fictional character.

2

Learning and Development

"I have to be honest with you, Andrea. I was completely shocked when he placed his resignation letter on my desk," John Riley said. A mug of steamy coffee sat neglected in front of him. "He was only with me for three months! The last person who turned over was at least with us for a year. Alexander barely gave it a chance."

John stirred his coffee absentmindedly, staring out the large window of Andrea's favorite diner. Andrea Alastor, president of Ala*star*, a large tech company just outside of Pittsburgh, stared back at her mentee. Before founding Ala*star*, Andrea had supervised John at two large tech companies, hiring him onto her team as a lead engineer both times. She was no rookie. With 25 years of experience in the world of technology, she ranked among the best in the country. When she met John, she was impressed by his work ethic and innovative spirit. It was just as baffling to her as it was to John that someone would leave John's company so quickly.

"Did he give any reasons for leaving in the resignation letter?" Andrea asked. "Three months seems pretty quick. He must have had some reason for wanting to cut ties so soon."

"The only thing he mentioned was that there was no room for innovative thinking," John replied.

> Which, Andrea, you know, is wrong. I pride myself on innovation and creativity. It's in my mission statement! It's my innovative valve designs that put my company on the map. I know he was struggling with one of my more recent designs, but I think he would have been able to figure out how to make it work if he had just stuck with it a little longer. I was waiting for him to finish up with this project so I could hand him another valve design I had been thinking about. Now I'm right back to being short-staffed.

DOI: 10.4324/9781003318354-3

John pushed his untouched coffee to the middle of the table and crossed his arms over his chest. He sunk into the booth as he reminded himself of Alexander's potential. He was exactly the kind of person they needed to push their company forward. Now, he was gone.

> I really thought I had something with Alexander. Once he had proven himself with the first few projects, I had planned to personally mentor him through my own innovation process. I could have made him a star at Trino-Tech. Once he learned the right way to approach innovation for our company, I know he would have made huge strides as an engineer,

John added.

"The right way?" Andrea prodded.

"Well, you know what I mean. The process I started at Omega Valves really helped set me apart from the other engineers," John tried to explain. "It's the same process that helped me get from Omega Valves to my own company. It's a process that works."

Andrea was shocked to hear John use phrases like "innovation process" and "the right way" as it pertained to innovative thinking. It occurred to Andrea that perhaps John embodied innovation, but only as it pertained to his individual abilities and contributions. It didn't seem like John was completely open to the idea of others contributing innovative ideas for the company.

"What kind of people do you aim to hire, John?" Andrea asked. She brought her mug of coffee to her lips, blowing gently before taking a small sip. She had a feeling they might be there a while.

"I like to hire young talent with an appetite for hard work and thinking outside the box. It's important to me that new hires have a work ethic that embodies both a creative and innovative process," John explained with pride. "I like to hire the people I think could one day be the best in the field and strive to develop new ideas for the company."

"That's exactly the kind of person I would expect you to hire; folks who are just like you," Andrea commented.

Now, when those folks offer up new ideas, how often are those ideas then implemented into the company dynamic?

"We are constantly adding new ideas to the company. It's how we've been able to stay competitive," John replied. "Just a few weeks ago, I developed a new valve design that will replace the one I designed two years ago."

"That's great, John, but I didn't ask about your ideas. I asked how often new ideas from other folks are factored into the workload." Andrea watched John's face turn from one of pride to one of discomfort.

John sat back in the booth and tried to think of the last time one of his employees had offered up an idea worthy of pursuing. Deliah did have a neat freeze protection valve design that they could have proposed to the new industrial complex builders, but it wasn't like anything they had tried before. What if it failed? Then there was Glen's idea for a new mixing valve, but it would have taken a lot of testing. At the end of the day, it was just easier to go with the valve designs that have always worked.

"I suppose there haven't been too many," John admitted with a sigh. "But my company is still so new. It doesn't seem safe for us to take too many risks quite yet."

"I'd like to tell you a story that might surprise you," Andrea offered.

> The way it starts might just sound familiar to you. As you know, I first started Ala*star* ten years ago. Not surprisingly, I was very similar to you. I was full of innovative ideas and only hired the most incredible talent I could find. The first year or so of my start-up went well. It went better than well; it was incredible. My company was growing at an unprecedented rate, and I attributed a lot of that success to myself. It was my ideas and designs that were getting us business, after all.

"Exactly! See, you get it!" John cut in.

> Not so fast. First of all, my ideas weren't the only thing making us successful; it was also the work ethic of that incredibly talented team I hired. Second, I thought that's all I needed to do; come up with ideas and watch other people implement them. What I failed to realize is that top talent from all over the industry also wants a way to implement their own ideas. They don't want to be stuck in a company that will never allow them to grow.

Andrea paused to let her words sink in.

I lost a lot of people after my first year, some of the best people I've ever worked with. I lost them because I was too afraid and naïve to recognize their own individual talent. I had an employee just like Alexander. He was fearless and driven. I thought the world of him. After just seven months at the company, he left, citing the fact that I refused to implement one of his ideas even after he was able to demonstrate its value. Just a little less than a year later, the company he left me for began distributing his design and it was one of the most successful products the company ran that year. I was devastated, and I decided right then that I wasn't going to keep losing good talent.

"I did some research and decided to instill and embody a culture of learning and development," Andrea finished.

"You mean like R&D?" John asked. "I don't really have the resources for that right now."

LEARNING AND DEVELOPMENT STRATEGIES

"Sure. Research and development is similar to Learning and Development. If you lack in resources for research and development then you lack in resources for innovation. They go hand in hand," Andrea jumped back in.

> **Many companies make the mistake of believing that they have to set aside million-dollar budgets for research and development teams, but that's simply not the case**. There are a number of strategies you can implement at your company immediately that won't cost you anything and, of course, a number of strategies you can implement over time as your budget changes.

"Okay, well, what did you start with?" John asked.

Developing Culture

"I started with the hardest question I've ever asked myself. **What do I want the culture of my company to look like?**" Andrea replied. She gauged John's reaction and saw a glimmer of confusion flash across his face.

> See, culture is more than just flexible hours and a good benefits package. Culture is something that anyone can walk into your company and *feel.* It's the way employees engage with one another and from the top down. It's the energy that exudes from offices and team meetings. It's the excitement and dedication people feel when they come into work for the day. I wanted a company culture that made people feel like they were contributing to something larger than themselves. I wanted my employees to come to work every day excited for whatever it was that waited for them.

"I get that. I think most companies want that, but you can't have that every day. No one loves coming into work every day," John pushed back.

> Of course, employees are going to feel stress, and not every aspect of every job is "fun", but it's not about maintaining excitement for paperwork. It's about maintaining excitement for the bigger picture, for the mission, and for the culture of the company they get to come to every day.

"Okay, so you asked yourself about culture. How did you achieve that culture from there?" John asked.

Well, I don't know that I ever truly achieved it, but it's something I work toward every day. The first thing I did was **invest in Lean thinking. I began with the Lean principle of creating a learning organization, where learning is at the very center of our daily routines**. For me, it was a substantial mindset shift from one of execution and relentless work ethic, to one focused on learning and development across the company. See, what I was failing to realize as the leader of my own company was that my ideas would only get the company so far. **What I needed was to diversify the ideas to make the company more agile and adaptable**.

> Innovative ideas don't come from just one person – I can't assume that I always know better than the people I've trusted enough to bring onto my team. In fact, quite the opposite has proven true over the past few years. Once I was able to create space for my teams to offer ideas freely, many of those ideas combined and compounded on one another, creating ideas that no one in the room would have ever been able to think up on their own. **It was the combined experience and brilliance of everyone involved that allowed the ideas to develop more easily**.

"Okay," John interrupted. "That sounds great. So, all I need to do is allow my people to toss around more ideas during meetings. I think I can do that."

"It's a bit more complicated than that. I also wanted my people to feel safe taking risks," Andrea continued.

"Safe? Why wouldn't someone feel safe?" John asked.

"Suppose you have been going to staff meetings for years, and each time you propose an idea, it's shot down with questions or garnering feedback. How do you think your mindset about proposing new ideas would change over time?" Andrea asked.

"I guess I probably stop proposing new ideas," John replied quietly.

Precisely. I noticed that my people had been contributing less during team meetings. So, I implemented monthly **think-tank meetings**. Before these meetings, I encourage any employee who wants to participate to bring a new idea to the table. They are the third Wednesday of each month, and I don't put any time limits on the meeting. The consistency allows my employees to plan other projects and deadlines around the meeting. During the meeting, each employee with an idea takes turns pitching their idea to the rest of the group. If the group seems to like the idea, we discuss its feasibility and allow the rest of the team to make improvement suggestions. Then, once the discussion is over, I give everyone a few days to reflect. At the end of the week, I send out an email for everyone to vote on ideas that we should pursue. Depending on the workload, we will run with one or two new ideas each month.

> These meetings have fundamentally changed the way my company operates. In these meetings, **every member of my team has the opportunity to interact with ideas and concepts they otherwise may never have thought of**. The **discussions are productive and positive**. Any criticism that is considered unproductive or insensitive is immediately addressed. Then, **allowing my team time to reflect** helps them think through goals and workload with a more level head, and I often see even more suggestions and ideas come out of this reflective time.

"Okay, but doesn't that take up a lot of time? I mean, how do you get any of the projects that really matter done?" John asked skeptically.

> I'm glad you asked. In a lot of ways, I think my company is more productive now than it ever was. By creating a culture that values collaboration and diversity of ideas, they have even more energy to take on new, exciting projects.

Experimentation

"I suppose I see what you're saying, but what about time? How do you balance the projects you know will work with a new project that you are unsure of?" John crossed his arms back over his chest as if he had just asked a mic drop question.

I think it's less about finding a perfect balance and more about considering my company's values. At Ala*star*, we very much value innovation and continuous improvement. **For my company to demonstrate those values, we have to be willing to take a few risks along the way. It's actually one of Lean's rules – to improve through experimentation**. It's not always easy – or balanced – but we've been able to continuously put strong products on the market.

"What if one of your risks doesn't pan out the way you thought it would? It seems like an awful lot of time, money, and resources for something that might not work out," John stared out the window. He realized that he was being more difficult than necessary. He was struggling with the notion that he was not only potentially wrong, but that he wasn't acting in accordance with his company motto.

"I'm glad you asked." Andrea remained calm. She completely understood why John was feeling so combative. He was losing strong employees without realizing that he was the one to blame. As that realization started to sink it, it made him feel more frustrated and inadequate. She knew these feelings all too well.

When one of our experiments fails, I try to remind myself that my company is not one that outputs 100% of every project it starts. My company is one that outputs only *the best* of *most* projects it starts. We don't push through and force something to work and then hope to sell it on the market. That's a recipe for disaster not only for the product but for the morale of my team. Instead, **we see failed experiments as a way to learn from what won't work and push us towards a more viable product**.

John continued to stare quietly out the window. He nodded along to what Andrea was saying, but she could tell he was caught up in his own thoughts.

"Can I tell you another story, John?" Andrea asked. John nodded his head and reluctantly turned to face Andrea.

Several years ago, right after I decided to start making changes to my company's culture, one of my employees came to me with a concern over the design for one of our voice-command device's soundboard. Now, I had personally redesigned the soundboard myself and was excited for its creation and implementation into the device, especially as we approached the pivotal holiday sales-period.

My employee came to my office and noted that while the new design should work in theory, there was likely a better way to approach the design for both efficiency and effectiveness. Feeling a bit put off by having one of my staff members so strongly critique my design, I turned down her proposal. "Make it work," I told her and sent her on her way.

A few weeks later, she came back to my office to let me know that she had attempted the design but was running into some issues during testing. It didn't seem like the circuit board I created could support the number of voice controls I had proposed for the project. I looked at her notes and the test results and realized that I had made a mistake. The circuit board needed to be bigger, which would mean a complete redesign. I checked the calendar and realized that there wasn't enough time for that. "We don't have enough time," I told her again. "Just find a way to get it done."

Now, I had just a few days before our meeting, given a speech to the company about innovation, and outlined the company's "new direction." Well, that employee took my speech to heart and ran with it. She collaborated with our software engineering department and was able to make adjustments to the design that didn't impact the timeline and only minimally impacted the budget. A few days after she did this, she came back to me and showed me the new design – this time almost completely her own. As you can imagine, I was completely shocked. "Find a way to get it done," did not mean "redo my design." I opened my mouth to let her know that I was unhappy with this move when she said something I will never forget.

> She said, "I was really struggling with this when you first told me to make it happen. I felt like we were forcing something to work and not looking at any other options. When you gave that speech about innovation, it opened up a whole new set of options for this project, and we found something that works even better!"

"What did you say to her?" John leaned into the table, hanging on Andrea's story.

"I said, 'sure thing!'" Andrea laughed loudly, and John chuckled along with her.

> Truthfully, I didn't know what to say. After she left, I looked over the solution she came up with, and it was brilliant. I mean, she worked hard to come up with something that would be cost-effective and stay on budget. It ended up

> being a sellout product. If we had stayed the course, I would have pushed through a product that would have tanked.

John contemplated Andrea's story. It wasn't unlike his situation with Alexander and reminded him to review the designs for the valve when he got back to the office.

Even though Andrea's story was compelling, John still wasn't completely convinced.

"It seems like this case only worked out because your employee was very skilled at her job. How can you trust that all of your employees have the skills to make those decisions and take those kinds of risks?" John asked.

Developing Skills

"That's a very good point," Andrea responded. "I think the answer to that question starts somewhere in the training process. What kind of onboarding and training do you offer your employees right now?"

John thought back to Alexander's onboarding. He recalled making an onboarding schedule for him, but then again, being a mechanical engineer in one place isn't that different from another. Why should he have to waste time and resources on heavy training when new hires should come in with the skills his company needs?

> Well, I have about a week-long onboarding process where the new hire is introduced to the rest of the staff, and someone brings them up to speed on the policies. The training itself is all on-the-job. They start with small projects and work their way up,

John reflected.

"**Do all of your employees come to your company having all of the basic skills needed to be successful?**" Andrea asked.

John stared at Andrea blankly. Of course, they did. How else would they have been hired? Even for Andrea, this was a strange question.

Andrea began to giggle.

> John, **I don't mean "do your employees have the *potential* to be successful at your company"**. Everyone at your company likely has that potential. **I mean, do they possess the skills**. Do they know how to use all of your

> software programs? Do they know how to communicate with customers or members of the production team? Do they know how to write up technical instructions for their products?

John could feel his blood pressure rising with each of Andrea's questions. If he was quite honest, he hadn't considered these questions before. Why would he? In John's eyes, everyone in this profession should know how to use software programs or at least figure them out. Communication and writing skills? Well, on second thought, he did find himself getting quite frustrated with Jim's long, rambling emails. He hadn't thought to check on his technical instructions.

"I'm not sure I quite understand this line of questioning," John admitted. He didn't want to sound annoyed with Andrea's prodding but was struggling to identify her intent.

"One of the biggest mistakes I made as a young business owner was assuming that all of my employees knew how to execute all of the basic skills required for the job," Andrea explained.

"Shouldn't that be a safe assumption?" John pushed back.

> Well, in the sense that they can write an email or struggle through a software program, yes, but that's not what I'm getting at. I'm talking about being able to do something *well*. **For example, not everyone is capable of writing technical copy in a clear, succinct way. Sure, many people can write the copy, if assigned the task, but not everyone can do it *well*,**

Andrea continued. "**And, if you want technical copy that production managers and clients can understand, the level of 'basic' knowledge needed is much larger than simply knowing whether or not someone *can* write**. Does that make sense?"

> I think I see what you're saying. Basic knowledge can mean different things to different people. Basic understanding of computers means one thing to a non-profit organization hiring a Director of Operations and an IT company hiring a Director of Operations,

John reasoned.

"Exactly," Andrea smiled widely.

"So, what does this have to do with training, then?" John asked.

FIGURE 2.1
We already know what we need

> Right. **One of Lean's principles is to observe work as activities, connections, and flows to gain a complete understanding of how work is completed on any given project**. In my opinion, to understand how work is completed, you must first understand how folks are trained. Are you familiar with the **70-20-10 Model for Learning and Development**?

John shook his head.

> We know that everyone learns new things differently, but **at an optimal level, people garner seventy percent of their knowledge from on-the-job experience**. This largely sounds like what your company does for training. They learn by doing, which is great because that covers most of how people garner knowledge. However, that's not all of it. **People obtain an additional twenty percent of their knowledge from their interactions with others**. Now, this can have both benefits and drawbacks as people will pick up the

> habits of others on their work team. If you have an employee that is always on time and eager to work on new projects, the employee in training is more likely to be on time and express eagerness. If that same employee has a bad attitude, then the new employee is equally likely to develop a bad attitude.

John racked his brain, trying to think about the people Alex was working with. Did any of them share similar desires to leave?

"**The last ten percent of knowledge comes from some kind of formal education**. This can come in the form of workshops or classes, but can also include formal meetings to review policies and procedures," Andrea said.

> **The point I am trying to make here is that all three components of learning are crucial to your employee's development**. If you are only relying on one or two, then you are missing the opportunity to get more out of the folks you hire.

John thought back to Alexander's departure from his company. While he had given him a handful of projects to work on, he hadn't paired him up with anyone he could learn from easily, and he certainly hadn't offered any workshops for Alexander to learn basic skills for his role. Perhaps John wasn't as mindful of professional development as he had hoped.

"That makes a lot of sense," John ordered another cup of coffee, this time planning to drink it while it was still warm and leaned into the table. "What kinds of formal education does your company offer?"

"It depends on the employee," Andrea responded, "but we offer a variety of opportunities." For Andrea, **professional development was unique to each of her employees**. Sure, **she has a standard training that each employee goes through to establish agreement of both what and how things should be done, yet another Lean principle**. However, she found that she was instead creating cookie-cutter robots instead of well-developed employees when she had tried to standardize beyond that. Instead, Andrea took a unique approach.

> For starters, I created an on-line learning account for all of my employees to access. This website allows them to take asynchronous classes on topics they find most interesting to their professional development. I set aside a small professional development budget for each of my employees to attend conferences and workshops of their choice each year. For client-specific learning, I am always sure to hold formal meetings to review client needs, wants, and

any specific things the team should know that might affect our relationship with that client. It's very robust but worthwhile. Since implementing these strategies in the last five years, my retention rates have soared.

John let out a long sigh.

"I'm not sure that I have the budget for all of that right now."

John, it took me years to have the ability to do this for my employees, but I knew I had to start somewhere. If the budget is tight, then I would consider offering professional development PTO. This allows your employees to take off from work for professional development without having to worry about tapping into their vacation days. It directly benefits you and the employee even if you can't pay for the experience they are hoping for.

"It's something to consider," John pondered.

After hearing everything you just said, however, I am a bit worried about my employees' interactions with one another. Twenty percent is a large percentage. How do I make sure that my new employees aren't picking up bad habits or negative vibes from others?

Continued Mentoring and Coaching

"Have you ever considered mentoring or coaching programs?" Andrea asked.

"I haven't," John replied. "Aren't they the same thing?"

Good question. Let's start at the beginning. First, assess your new employee's needs and determine whether or not **mentoring or coaching** are viable options for them. This will ensure that your employee is being paired with a mentor or coach that you trust to provide positive experiences and habits. **This also helps to enhance your learning organization**. The other way is to address negative habits and vibes head-on, which requires taking a strong look at your **leadership practices**,

Andrea replied.

Let's think about mentoring first. Mentoring is the most obvious and can also be the most rewarding for the employees at a company. **Mentoring is a**

> **great way to keep more senior-level employees active and engaged in the development of the company and it shows that you are invested in your new employees' development as well.**

"Sort of like you did for me back when we worked at Fortune One," John reasoned.

> Exactly. And sort of like I do for you now. Essentially, **the focus of mentoring is to help develop a well-rounded mentee by offering support and general knowledge area expertise that the mentee can learn and grow from. These programs are known to have lasting effects, including employee retention.**

"Okay, let's say I wanted to start a mentor program at my company, how would I go about doing that?"

"Great question, but you're not going to like the answer," Andrea joked.

"Let me guess: it depends?"

FIGURE 2.2
Learn to improve

"Exactly," Andrea laughed.

But there are tons of great examples for what companies like yours are doing with their programs. Caterpillar, for example, pairs new employees with a mentor for the first three years of their job and allows employees to rotate to different departments. As you can imagine, allowing people to rotate through other departments is a catalyst for innovative ideas. General Electric does a pretty unique mentoring program they call "reverse-mentoring" where young employees mentor more senior employees on things like technology or generationally-unique perspectives. Schneider Electric uses an AI system to match mentees with mentors based on their goals. **The main thing to keep in mind is that you don't want to arbitrarily assign mentees to mentors. It's important to be thoughtful about this process and pair folks with similar styles, goals, aspirations**. Otherwise, it's really up to you to decide how you want to set up a mentoring program that will work for you and your company's needs.

John sank back in his booth and began carefully thinking through all of the things that Andrea had mentioned. Out of all of them, a mentoring program seemed to be the most immediately feasible. Everything else, though? He felt like his brain was swimming in choppy waters. How would he ever be able to sort through all of this?

"Deep in thought, I see," Andrea commented.

"Well, you've given me a lot to think about."

"Perhaps too much, it seems," Andrea chuckled.

"A lot, but not too much yet," John laughed with her.

I think you mentioned something about coaching. I've heard about executive coaching before. Mason, one of the Directors at Fortune One, just hired an executive coach for himself, and he said it's been extremely helpful, but I'm not sure I quite understand how coaching is any different than mentoring.

The first difference to note is that coaching is more of a short-term relationship with a specifically **designed goal**. For example, a mentor might let you know that you should try out a mentorship program at your company, as I've already done today. After leaving here then, you might hire a coach to help you design an effective mentorship program. It's a subtle difference, but an important one as mentors are here to provide long term guidance about a plethora of topics, while coaches hone in on just one or two topics.

A second difference is in its structure. Our mentor–mentee relationship is more informal. We meet when something comes up for you or when you just want to get together and chat. **A coaching relationship comes with structured meetings predetermined by both the coach and coached.**

> **Another difference is that coaching is designed around performance indicators and improving skills. It is important to note that coaching relationships have a specific and measurable outcome around those performance indicators or skills.** For example, if you were working with a coach on your mentoring program, you would define that goal by setting specific key indicators for how it will be accomplished along with a way to measure its success, perhaps through retention over three months or through employee-satisfaction surveys. Does that make sense?
>
> Yes, that makes a lot of sense, but all of this is starting to add up. While Mason did mention that his coach has been pivotal in helping him develop, he also mentioned how expensive it was to retain his coach. I'm not sure the company could afford a coach for all new employees,

John replied.

> I'm really glad you mentioned that. **While on-the-job-coaching can be an extremely valuable asset to any company, it really shouldn't be a requirement for every new employee. Coaching should be used when you have a more specific, performance-driven goal or issue you want a particular employee to work on.** Not to mention, there are many different kinds of coaching, like skills coaching, performance coaching, and even group coaching. The key in all of this is to identify what kind of coaching your employee needs, if any, and then assess how you might go about implementing a coaching program.

"You mean hiring someone to provide coaching?" John asked.

"Not necessarily," Andrea replied,

> but I wouldn't rule it out either. Honestly, John, if you think coaching is something your company might want to look into, I highly recommend hiring one or two coaches for your company or consider sending two of your current employees who might have an appetite for this kind of work through a coaching program. While this might sound like an investment up front, I can almost guarantee that it would pay off in the long-run as you find the benefits of coaching more profitable.

John nodded his head and took another gulp of his coffee.

Listen, I know I threw a lot at you just now, but **both coaching and mentoring can be a breeding ground for innovation**. Senior-level mentors offer knowledge based on their years of experience, and mentees put a new, updated spin on that knowledge, helping to develop something completely new. Coaches are highly trained to ask incredibly thought-provoking questions of their clients. It is a coach's job to help their clients see things from a different angle, which often inspires new ways of thinking through and about problems they are facing.

> I know how much you value innovation and creativity in your world. **Give your folks the ability to collaborate and discuss ideas with one another more often and more consistently and it will give you the innovation I know you're looking for**,

Andrea finished.

Adequate Leadership

> You've mentioned a lot of things today that I have no doubt will help, but I'm not sure where I'm going to get the money to afford all of these programs. I know mentoring might not cost me anything upfront, but coaching, professional development, even experimentation could cost the company a lot of money. I'm not sure how I'm supposed to make all of this work.

John stared back at Andrea blankly.

"I hear you. I know budget isn't always where you want it to be when trying to make these changes. Trust me; I know this well, which is why I saved the free strategy for last," Andrea smiled.

"Free? As in no cost whatsoever?" John asked for clarification.

"Well, it depends on how you view cost. See, **this strategy focuses on how the leader promotes the culture we discussed at the beginning of our conversation**," Andrea began.

> Essentially, this strategy requires you to hold a mirror up to yourself and ask, "Am I promoting and contributing to the culture I want to create?" Now, this might sound similar to the first question we asked today, but **there is a fundamental difference between knowing what kind of culture you want and empowering others to create it versus actively participating in its creation yourself**.

"But it sounded like you did that with those meetings. You even said they were successful," John countered.

I'm glad you brought those up! Those meetings did change our processes, and they did work for a while. I would say after about four months into starting those meetings, I noticed that fewer and fewer people were bringing in new ideas. At one of the meetings, only one person presented, and she looked squeamish about it. She even began her presentation with "I'm not sure if this is a good idea, but…" I grew worried. How could something that was going so well suddenly fall apart?

> I thought I was doing everything right. I stopped being overzealous with my ideas at meetings, and I just allowed myself to sit back and listen to my people. Their ideas were excellent, and I learned a lot about my employees in ways I never would have before,

Andrea continued.

"Okay, so what happened?" John asked eagerly.

> Well, I did stop proposing my ideas, but that didn't stop me from moving forward with them anyway. I had done everything I could to empower my people to create a process for which new and innovative ideas would be proposed and assessed equally, but I didn't stop for one second and consider whether or not those rules should apply to me as well,

Andrea answered.

"Well, why would they? You're the owner of the company. Why should your ideas have to be vetted like everyone else's?" John asked.

> See, I thought those same thoughts, but the real question is, "why shouldn't they go through that process?" **Every single idea proposed at one of those meetings came in looking good but left looking great or phenomenal**. Was I so great that all of my ideas just came out phenomenal? Of course not. Not to mention, pushing my ideas through also meant that someone whose idea had been properly vetted took a back seat to the one that came from my office. It was demoralizing to my people.
>
> I can understand how that makes sense, but I'm not sure I agree that *all* of your ideas need to be vetted through the same process. I can see how it would be upsetting if none of them were vetted, but do all of them *need* to go through that process?

"Another great question!" Andrea's face lit up with a wide smile as John continued to travel down the same thought process she used to justify to herself just a few years ago.

> I thought the exact same thing. I started to propose a few small ideas at the meetings again. I did my best to keep it to a respectable minimum so others would have a chance to propose their ideas. It took a few weeks, but things started to turn back around. I saw more participation in the meetings, and my folks seemed happy again. That is until one of my ideas hit a snag in production. The project manager in charge of my idea brought it up at one of our meetings, and I was dumbfounded. Sure, we openly discuss minor failures and hiccups of other projects but to discuss an issue with one of my projects? I couldn't believe it. Instead of answering her question, I told her that she should address questions of that nature to me in private. The room went completely silent. It felt like pulling teeth to get anyone to talk the rest of the meeting.

"Again, you are the boss. It is inappropriate for them to bring this up in front of your other employees. I think you did the right thing," John praised.

> I thought so too. The project manager came into my office later that afternoon to discuss the production error. After a relatively brief conversation, we were able to work out the problem and send it back through testing. No harm, no foul. As she was leaving my office, she got visibly upset and began apologizing profusely. I was so startled by her change in demeanor and didn't want her to leave feeling upset. I asked her to explain what she was thinking, and she said, "It's just that you've always been so understanding when it comes to seeing failures as learning and growth opportunities. I didn't stop to think that issues with your projects might be viewed differently. I should have thought through that more, and I'm sorry for that."

"But why, Andrea? I don't understand. You *are* the boss. Your ideas *are* different," John emphasized.

> That's where you and I disagree. **If I truly wanted to create a culture of learning and development, which I very much believed in, then I had to be willing to fully immerse myself – and my ideas – into the process. It's not truly a learning organization if the leader isn't actively participating in the learning process.** While I was supporting all of these creative

> initiatives to bolster innovation with back slaps and handshakes out in the open, I was still doing things "my way" behind closed doors. I wasn't ready to let go of my comfort blanket and truly lean into the process that I was telling everyone I believed in, and it cost me a few more great people.

For the last time, John leaned back in the booth and let out a heavy sigh. It's not that he didn't believe Andrea's stories. She wasn't one to lie, but he still didn't see anything wrong with being seen as superior to his employees.

> I know what you're thinking, but this isn't about letting go of being "the boss" or the person who built a company from the ground up. You are the boss, and you did build a company from the ground up. That's impressive, and people will respect you for that all on its own, but I think people might respect you even more if you demonstrate the things you want from them. For me, I wanted that culture of learning and development, which means I had to be okay showing my folks that mistakes weren't fatal; that they were part of the learning process. It was a hard lesson to learn, but I know my company is better for it.
>
> I don't know. You gave me a lot of great advice today. I love what you said about helping my people develop their skills and investing in a stronger culture, but my company is still so young. It's not just budgeting issues that concern me, either. I need my people to respect me. Taking too many risks, especially if those risks result in failure, might ruin my credibility.

John slid to the edge of his booth uncomfortably, signaling his preparation to leave.

> Don't forget that I was once where you are now. It took me a long time to feel safe enough to start implementing any of what we discussed today. I just know that if I had done it sooner, my company would have grown much quicker.

Andrea offered her final piece of advice and let the conversation come to a gentle end. She didn't want to put too much pressure on John. She had given him enough to think about.

John glanced down at his watch to check the time. He had been with Andrea for much longer than he had anticipated, as was usually the case for the two of them. "Another two and a half hours, gone just like that," John said with a smile.

"It's always a pleasure, John," Andrea smiled. "If you need anything else, please don't hesitate to reach out."

John stood up from the booth and walked out of the diner into the blustering cold afternoon. He barely noticed the chill set in as he contemplated everything that Andrea had told him that day. Still, Andrea had never led him astray, and she always had his best intentions at heart.

"Alright, I know I need to do something *to keep people from leaving, and it sounds like it all starts with the culture,"* John thought as he made his way down the snowy sidewalk.

QUESTIONS TO CONSIDER

1. How do you support and contribute to learning and development at your company?
2. What do you want the culture of your company to look like?
3. How can you bring more diversity to the ideas proposed at your company?

FIGURE 2.3
Run Einstein run

4. What can you do to give your employees more time to think and discuss ideas?
5. What kinds of risks does your company take?
6. How could your onboarding process be improved?
7. How can you provide more unique opportunities for professional development?
8. How would your company benefit from a mentoring program?
9. If your company already has a mentoring program, how might you improve that program?
10. How often do you assess your leadership practices?
11. What more can you do to demonstrate support of the culture you are trying to create?
12. How do you actively participate in your company's culture?

3

Group Contributions to Innovation

Michelle Ward pulled her long, black coat tight around her body. The wind chill had dropped significantly since the sun went down, and she could feel the cold chatter through her teeth.

I can't believe Mary Lynn still wants to get together in this, Michelle grumbled to herself. It's got to be five below out here.

Michelle and Mary Lynn had been coming to Acacia since their college days. Back then, Acacia operated as more of an authentic speakeasy, the kind you had to give a password to get into. The password was always listed on their website, but it did add to the ambiance of the bar. Now, you could just walk right up to the door and let yourself in. The new owners lacked in innovative thought.

Mary Lynn was sitting at a large, plush booth in the back of Acacia. Her nose was buried in her phone, typing away at some unknown message.

"Had to find the booth furthest away from the door, huh Mare?" Michelle asked.

"No, furthest away from the cold breeze the floats through the door every time it's opened." Mary Lynn smiled widely. "Give me just one sec to finish up this email."

Mary Lynn stuck her tongue out just slightly, curling it around her upper lip as she furiously finished typing the email on her phone. This little quirk always made Michelle laugh. It was her tell: something exciting is afoot.

Mary Lynn struggled right out of college two years ago. She applied to countless jobs before she landed the job at Ala*star* just a few months ago. She did the same tongue curl the night that she told Michelle she had gotten the job.

Michelle braced herself. It's not that she didn't want to hear what Mary Lynn had to say, but she was having a horrible week at work.

DOI: 10.4324/9781003318354-4

TrinoTech hired Michelle right out of college. She was drawn to the owner's innovative attitude and loved the idea of being part of a start-up company. However, her excitement quickly dwindled when she started to realize John wasn't quite the champion of innovation he had made himself out to be.

"What's on your mind?" Mary Lynn asked, breaking Michelle from her thoughts.

"Oh, it's nothing," Michelle responded. "What's going on with that email of yours?"

"I can tell it's not nothing." Mary Lynn ignored Michelle's question. "Are you still having trouble at work?"

> Not much has changed, so yes. I tried to propose another idea today for a new heat control, but John completely shut me down in front of my entire team. He didn't even look at the design I came up with,

Michelle vented. "Maybe he doesn't think any of my ideas are worth looking at."

"What are you talking about? You're one of the most innovative people I know! It sounds like he's just not a very innovative leader," Mary Lynn consoled.

KEY CHARACTERISTICS OF INNOVATIVE PEOPLE

"What makes you think I'm so innovative?" Michelle shot back. She could still feel the frustration from her rejection at work.

Creativity

"I don't know, Michelle, how about the fact that you are one of the most **creative** people I know?" Mary Lynn began.

> Do you remember Professor Simpson's robotic arm? He couldn't figure out a way to make the movements fluid. He had tried everything. He even offered to give anyone who solved his problem extra credit. No one could figure it out.

Michelle smiled and nodded. She knew what was coming next.

> You poured over the problem for weeks: gears, oil, too many parts, not enough parts. You were **always contributing new ideas and suggestions** to make the movements more fluid. You **helped build a creative environment** for the rest of us to contribute new ideas and potential drawbacks. Each time someone brought an idea to the table, **you quickly identified potential problems or drawbacks**. If the idea had potential, you helped us push through the **creative process**. You would help everyone **prepare** by **jumping in with the research and gathering additional information**. Then, you would **encourage everyone to sleep on it for the next few days to let the idea incubate. Sometimes you were the one who experienced the epiphany and illuminated the rest of us, but you were equally excited and supportive when others were the ones generating ideas**. You were hands-on during the **evaluation stage** and always **made sure that the ideas were analyzed properly**. You never left an experiment unfinished, even if you were convinced it wouldn't work. You loved the idea of learning new things from concepts that didn't pan out the way we thought they would. And then there is always that final day of experimentation. Do you remember?

"Oh, I remember," Michelle replied. She wrapped both hands around her drink in humble memory.

> It was your idea we were working through, and it passed all of our tests. We took the plan to Professor Simpson, and he let you **implement your idea** onto his robotic arm. I'll never forget the hoots and high fives from the guys in the class who desperately needed that extra credit,

Mary Lynn laughed.

> You approached everything in life just like that. **You were always curious, and you couldn't stand the thought of passing up a good challenge**. Not to mention the fact that you're a self-starter. I still can't wrap my head around the fact that you used to get up at five a.m. just to catch up on the latest news in the field. Your **intrinsic motivation** is unmatched when it comes to the creativity game.

Michelle was grateful for the memories. They reminded her why she got into the field in the first place. Mary Lynn was right; she was a hard-working self-starter, but maybe the problems were more in what she did after she got an idea off the ground.

"I wonder if my creativity isn't enough for John, though," Michelle replied. "Maybe I'm just creative, and I haven't taken that leap into innovative yet."

Note Taking

"I find that hard to believe. Don't you remember all of the innovators we learned about in college? You seem to emulate just about everything it means to be an innovative person," Mary Lynn replied.

"If you're going to bring up my journal again, then you're going to pay for the next round," Michelle joked. Since Michelle was young, **she carried around a small journal in which she kept all of her notes. She didn't know that it was one of the key characteristics of an innovator until a professor mentioned it during one of her college classes**. Michelle was an observant person who found just about everything interesting. She chose to journal, so she didn't forget any experiences or ideas from the day.

Opportunistic Mindset

"No, I'm not talking about your journal, although that is one of the characteristics of a good innovator." Mary Lynn rolled her eyes. "I'm talking about your weird ability to turn everything into an opportunity."

"What do you mean by that?" Michelle asked.

> Struggling to find a job after college was one of the hardest things I ever went through. I felt worthless, useless, and tired of missing opportunities I felt I was more than qualified for. Each time I came to you with another rejection, you helped me pick myself back up without giving me cliché lines like "it wasn't meant to be", or "you'll get 'em next time". You took the time to study the market, and you created a spreadsheet for me, outlining all of the companies that were hiring and what you thought their pros and cons were. From your research, you were **able to detect patterns in the market** that I would never have seen and helped me identify the right companies to apply for. And you made it all look fun!

"Well, I do tend to **crave complex experiences**. Putting together a job matrix for you was about as complicated as it got," Michelle joked. "But to be serious, a lot of those companies didn't pan out."

> True, but I still had more options to apply for than I would have had on my own. Plus, that's not all you did for me. I was struggling financially as soon as my loans hit. I had been looking vigorously for a job for six months and had nothing to show for it but was still expected to pay back all of my loans. I came to you in a panic, but you told me not to worry. You sat down with me and helped me strategically apply to a bunch of part-time jobs and volunteer work. The kicker?

"I made sure those part-time jobs and non-profits had connections to the companies you were hoping to apply for," Michelle admitted.

"And so they did," Mary Lynn agreed,

> big ones! The lead dog walker at the animal rescue was the Director of Engineering at Ala*star*. **With just a few pieces of random data, you were able to extrapolate out these patterns for creating opportunities in a systematic way**. I would never have thought to explore those patterns. See, you've **always gone above and beyond** to make things work both for yourself and others.

Values Social Capital

"Well, I think **there's also something to be said about the value of people**. The more people you know, the more likely you are to both learn from them and make connections you might need down the road," Michelle reasoned. "I used my networks to help me get interviews before as well. Knowing people is important."

"And I wouldn't have my job without the people you knew."

Mary Lynn did have a point. When Michelle was young, her mother told her that people were important. The more people you know, the more well-rounded you are, and the more resources you have to help you accomplish good things in this world. Michelle had taken that to heart and had always strived to grow and develop a diverse network.

Even when she started her role with TrinoTech, she **made a point to set up short meetings with everyone in the office to learn a little more about who they were and their roles at the company**. As usual, she took notes of particular things they mentioned, like birthdays or anniversaries. She also noted if that person mentioned any particular area of expertise they possessed or any areas of growth they mentioned wanting. If she had an interest in their expertise area, she mentioned wanting to learn more from them. If she had particular expertise in an area they mentioned wanting growth, she offered to help. It was simply the way she built her network, and it never failed her.

This was particularly useful just a few weeks after she was hired. Her officemate, Mikel, had gotten negative feedback from John about a design he was working on. Mikel was originally given the project because of his expertise in simple valve mechanics. Then, John decided that he wanted to add a mini-control panel to the outside of the valve to read temperature. He also wanted it to push notifications to a user's phone if the temperatures were too hot or too cold. However, Mikel did not have much experience in software development, which was required for this kind of project.

Michelle didn't hesitate. She went back through her journal and quickly identified two people who mentioned having a lot of experience with software design. She connected Mikel with the two of them. She also stuck around to help Mikel see the project through, offering her services as a sounding board for his ideas and how best to propose them to the software experts. Together, the four of them developed an updated valve control that not only provided temperature readings but flow readings as well.

As brilliant as Mikel was, he never would have been able to accomplish that project without the help of Michelle and the two software engineers. **It truly was an innovative idea, and Michelle recognized that innovation was the product of teams.**

Proactive and Persistent

"It's just like the charity event you organized in college." Mary Lynn broke Michelle from her concentration. "I don't know if I've ever seen anyone act as **proactively** as you did for that event."

When Michelle was in college, she organized a charity event to encourage adoptions of orphaned pets. The idea came to her after finding a tiny kitten outside her residence hall. It was a cold Pittsburg night, and Michelle decided to bring the cat back to her room. The next day, her resident assistant (RA) found it and told her she couldn't have it in the residence hall. Luckily, she had a large network of friends and family who were able to find a perfect home for the little guy.

Michelle was happy she could rehome him, but she wanted to do more to help other animals. She couldn't stand the thought that there were tons of other animals left out in the cold every night. So, she came up with an idea she called Adopt a Human on Senior Day. Essentially, orphaned animals were brought to campus during Pitt's Senior Day. Pitt seniors would then have the opportunity to "be adopted" for the day by one of the pets. The pets would "walk" the seniors and "play" with the seniors. The goal was for a graduating senior to fall in love with the pet and adopt him or her. The only problem was that it required the collaboration of a few different animal rescues, a ton of manpower, and even more planning hours to organize all of it.

However, Michelle was **not easily discouraged**. Even when the university pushed back on the idea for liability purposes, **she persisted**. She wrote a letter signed by over one-thousand members of the campus community and added a few animal bios and sent it to the board of directors. They were so impressed (and moved) by her letter that they allowed the event (and one of the board members adopted one of the dogs in the profile).

She reached out to a core group of people who began planning and organizing while another group of people helped with donations. They adopted out 15 pets in the first year. They doubled that number the following year. That was her sophomore year of college. Since then, the program has adopted out hundreds of stray animals to graduating college students.

> That project was no small feat, but you kept on persisting. **You never stopped, no matter how many obstacles were thrown in your way, and you made it happen**. My little sister said it's still one of the most popular events of Senior Week now. All because you wanted to be proactive and rescue more kittens,

Mary Lynn laughed.

"Okay, okay. Yes, proactive and persistent do sound like they describe me in that situation," Michelle agreed.

Challenges Status Quo

> **Innovative people challenge the status quo**. They don't accept that things can only be done one way. They look for several ways to do things, and **they question everything**. That's what you did with the kitten. I've never known you to wait for something to break to make it better. You **naturally look for ways to make things better all the time**. That's why I think this job situation is bothering you so much. Your boss isn't giving you the space to do things you naturally do on your own,

Mary Lynn explained.

> It's just so hard to see you so upset. You have always been the person to **push back when you think something is wrong and refuse to take no for an answer**. It seems like he is silencing that part of you.

Michelle took another sip of her drink and set it down on the table. She hated the thought that she was performing as anything less than herself.

"I do think there is one area of innovation you could probably work on," Mary Lynn said carefully. "You're always good at challenging other people, but maybe this time, you need to challenge yourself. I'm worried that you are grappling with some kind of limiting belief that's holding you back."

Michelle struggled with her friend's words for a moment, but she knew that Mary Lynn was right. **Innovative people did tend to look inward and assess their personal beliefs**. It was how they stayed on top of their personal growth internally.

"What do you think is holding you back?" Mary Lynn asked.

"I'm worried that I'm starting to believe that I'm not cut out for this company," Michelle said after a few silent moments of thought. "I think I know deep down that this belief is wrong, but I haven't been able to shake the feeling for a few weeks now."

"How can you challenge that belief?" Mary Lynn asked.

"I think I need to try a new approach. Maybe asking for some professional development will help demonstrate my commitment to being a strong employee," Michelle responded.

"I like it! Just another way for you to challenge the status quo at work," Mary Lynn praised.

GROUP CONTRIBUTIONS

"Funny you should mention professional development, though," Mary Lynn added. "That email I was working on when you got here was to follow up about a conference my company is sending me to."

"That's awesome! What kind of conference?" Michelle asked.

"It's a Lean conference. Lean includes things like streamlining processes, building a learning organization and eliminating anything that seems wasteful. I'm hoping to help my company identify more areas where we can create value for our customers," Mary Lynn explained.

"That's awesome! How did you find out about that conference?" Michelle asked.

"Our company has been expanding. Andrea opened up two new departments to focus on other areas of engineering. At first, it was great, but we started noticing some issues with things like communication and process management," Mary Lynn began explaining.

"I'm sure that drove you crazy," Michelle commented. "You are big on order and organization."

"Exactly! I took a tip out of your book and got myself a journal," Mary Lynn admitted.

> I started taking notes about the kinds of communication that were problematic and where in the process we were all getting stuck. I noticed a few patterns and took my notes to my supervisor. He loved that I was keeping track of the new departments but encouraged me to think about how some of the issues I was finding might best be resolved. So, I set up a meeting with one person from each of the new departments. It turns out that they were both tracking issues on their ends and we were able to come up with a few solutions to propose to our respective supervisors. Long story short, my supervisor told Andrea, the owner, about my taking initiative, and she pulled me into her office. One of her best friends, Marcus Jiminez, is a Lean expert who focuses on developing solutions similar to the ones I was proposing. He is hosting a Lean conference in Pittsburgh for people in our line of work, and she offered to send me to the conference.

"That sounds amazing! Your boss seems like she's very supportive," Michelle commented.

> Well, she's been pushing a few new initiatives lately because of the new departments, and she's been working hard to create cohesion even though we are getting larger and more spread out. A few weeks ago, we had a company-wide one-day mini-conference. She opened the conference with a keynote about the **vision for the company**. It wasn't one of those fluffy speeches you sometimes hear about or watch on TV, either. This one was **compelling and wildly motivating**. Our supervisors told us about our team objectives for the next two quarters, but Andrea **explained how each team objective supported the other teams objectives**. It **provided meaning** to what can sometimes feel like team-specific objectives.

"That sounds amazing," Michelle said.

> The last time John gave a speech, he simply went over the numbers for the last quarter. After that, he congratulated one of the teams for finishing a product that both my team and the software team helped make. He then finished with a cheesy quote to "believe in ourselves" as a way to motivate us to get our numbers up in the next quarter.
>
> Your boss could take a lesson from mine. Andrea is good at creating **goal interdependence** at our company. **Even if my team has a goal or objective we need to meet we know that our piece's success is connected to another team's success.** It **helps us be mindful of our collaborations with others.** We are constantly asking ourselves, "how can we complete our work on this project to set up the next team for success?" Other teams ask themselves the same question. It helps us with our **task orientation.** If **everyone at the company holds themselves to the same high standards, then we are more likely to create strong products for the market.**

"That makes a lot of sense. It sounds like everyone at your company also really enjoys working together to a certain degree," Michelle said.

> I like to think so. Of course, you might not love everyone you work with, but Andrea does a good job making us feel like **one cohesive team. Even if there are times when folks may not like one another on a personal level, everyone has the same commitment to the outcome, and that's what keeps us moving forward.** Andrea places a lot of importance on both **internal and external communications** at the company. My team has **strong internal communication that allows us to work efficiently and effectively, even if one of us is having an "off" day.** I think its part of the reason Andrea got excited when she saw that I was working so closely with the new departments. She loves to see cross-team collaboration. She knows that **external communication, when done well, can bolster innovation since the goal is to learn from and grow with one another,**

Mary Lynn explained. "I think that's why she was eager to recommend I go to the Lean conference with her."

"I love that your company understands the need for continuous improvement. It sounds like Andrea is a **champion for innovation. She doesn't just say that she wants to be innovative; she actively seeks innovative methodologies,**" Michelle replied.

"That's a fair assessment. She **expects innovation to be a part of our everyday process, and she is very supportive of risk-taking even when it doesn't go well.**"

"She sounds incredible, and so does this conference. I wish John would send me to something like that. That's just the thing that might help get our company moving in the right direction," Michelle said.

> Then go! This could be the professional development experience we were just talking about. You can use it to challenge your own limiting beliefs and the status quo at your company. The worst thing your boss can say is "no". It can't hurt to ask,

Mary Lynn encouraged. "Lean practices are the future of our industry. A conference like this could help you stand out on a resume if you want to leave your company."

Mary Lynn was right. A Lean conference like this could open up so many doors for Michelle and her future. There was no reason for her not to go. She decided to put in the request for professional development the next day.

QUESTIONS TO CONSIDER

1. Do you nurture the creative process with the folks you work with?
2. Do you keep a journal?
3. How often do you review your notes?
4. How do you identify opportunities in your line of work?
5. How often do you check in with your network?
6. How often do you make an effort to add to your network?
7. How do you show people in your network that you care about them?
8. When is the last time you challenged the status quo?
9. How do you challenge your limiting beliefs?
10. How often do you review the vision of your company with your teams?
11. What do you do to motivate your teams to work collaboratively with one another?
12. How do you create goal interdependence at your company?

4

Using Lean Principles and Rules to Foster Innovation

"Michelle! Over here!" Mary Lynn waved from the far side of the lobby. They had arrived 45 minutes early, and the conference center was already buzzing.

"Popular conference," Michelle commented.

"Marcus is one of the leading Lean strategists in our industry," Mary Lynn replied. "I still can't believe you were able to get in! They said it sold out just a few days after we last met."

"Someone must have dropped out." Michelle shrugged her shoulders. "I'm just glad I'm here!"

It was sheer luck that Michelle was able to attend the conference in the first place. Two days after she met with Mary Lynn, John announced that he was developing a learning and development team for all TrinoTech employees. The email Michelle received from her supervisor informed her that each employee could request a small stipend to participate in continuous learning opportunities throughout the year. It went into effect immediately. As soon as Michelle finished reading the small print, she ran to her supervisor's office to request a stipend for the Lean Strategies Conference.

"Michelle, I'd like you to meet Andrea," Mary Lynn said as Andrea approached the two women.

"It's a pleasure to meet you," Michelle shook Andrea's hand. "I've heard so many incredible things about your company."

"Thank you," Andrea replied. "My company is only as good as my employees, so I like to think it is pretty incredible. What do you do?"

"I am an engineer at TrinoTech. I've been there for about a year now," Michelle responded.

"Ah, so you work with John. Is he here as well?" Andrea looked around.

DOI: 10.4324/9781003318354-5

"I don't think so. He just recently opened up a learning and development budget for our whole company and encouraged all of us to take advantage of professional development opportunities. That's the only reason I'm here today," Michelle explained.

"Did he now?" Andrea smiled. She knew that their conversation had some impact on John, but she wasn't necessarily expecting him to implement everything so quickly. "Well, I'm very glad to hear that. John is a good man with a lot on his plate. I'm sure that was a big step for him to take."

Michelle only nodded in response. If Andrea knew John, Michelle didn't want to say anything that might get back to him.

"Look, there's Marcus now." Andrea walked a few feet towards the entrance to the Grand Ballroom to greet Marcus Jiminez. Marcus was the owner of HydroTherm, a hydraulic valve company, as well as the keynote speaker for the Lean Strategies Conference.

"Hello, Andrea." Marcus held out his hand in greeting. "You are looking well. I'd like to introduce to you one of my newest hires. This is Alexander. Alexander, meet Andrea."

"It's nice to meet you, Andrea. Marcus has told me a lot about you," Alexander said.

"It's a pleasure to meet you as well." Andrea turned to introduce the two women she had in tow. "This is Mary Lynn, one of mine, and this is Michelle. She currently works for TrinoTech."

"I thought you looked familiar," Alexander commented, looking at Michelle. "I worked there for a few months before joining Marcus's team."

"That's right. You worked with Deliah for a while. They haven't found a replacement for you yet. I know that team misses you," Michelle commented.

"It's a small world, for sure," Andrea noted. She looked over Alexander's confident demeanor and immediately understood why he couldn't work under John's leadership. At least, he couldn't do so the way it was now, but she was certain things would change.

"So, what have you cooked up for us today?" Andrea asked, returning her attention to Marcus.

"Well, if I told you now, it wouldn't be much of a surprise, would it?" Marcus joked. "Honestly, my job is only to get you warmed up. The real fun takes place in the workshops. Make sure to take advantage of them."

Marcus gestured for the group to find their seats so the conference could begin.

THE MYTH OF LEAN AND INNOVATION

"Good morning, ladies and gentlemen," Marcus began his keynote speech. "How is everyone feeling today?"

A muffled roar of collective "goods," "greats," and general groans of acknowledgment rose from the crowd.

"Well, that doesn't sound very promising," Marcus joked. "Hopefully, you find a bit more excitement throughout this keynote and your day. Also, there's coffee in the lobby for anyone who didn't see it."

A few folks laughed as Marcus continued his speech.

> As many of you are quite aware, this is a conference about Lean's role in innovation. How does it contribute? Why is it important? How can I start to implement it? These are all very important questions, but I think it's important to start by addressing the myth of Lean and innovation. As a Lean specialist for almost two decades, one of the first questions an executive will ask after their initial consultation is "won't these rigid Lean practices stymie the innovation of my team?"

Marcus paused to look around the room, taking in the number of people who nodded along with his introduction.

The first few times I heard this question, I felt caught off guard. I couldn't understand how folks were missing the link between Lean and innovation. The more meetings I had, however, and the more clients I took on, the more I realized that part of the problem is how to define innovation. See, most people have a poetic understanding of what it means to be innovative. When asked about innovators, many people point to Steve Jobs or Bill Gates – the lone wolves of innovation. Though, we will discuss in one of the workshops today how the concept of the lone wolf is, in and of itself, a problematic way of defining innovation.

> Bill Gates and Steve Jobs are certainly innovators. There is no doubt about it. And because they are the most obvious candidates for embodiments of innovation, many folks picture a young man sitting at an old table in his parent's garage. He furiously scribbles ideas down on a page before balling them up and throwing them in a trash can over and over again until he comes up with that one brilliant idea.

Several folks in the crowd laughed at Marcus's joke.

> What I began to realize is that **many people think of innovation as more of a chaotic or haphazard process by which one person comes to a eureka moment**. Now, I'm sure this isn't the first time you're hearing about this myth, but for some reason, it is still being perpetuated in the workplace. See, **the folks who ask me if Lean will deter folks from innovation are the same folks who think that order detracts from creativity**. It is the knowledge that too much structure and too much rigidity impede creativity, but not knowing where to draw the line. So, when I walk into their office and review the rules and principles of Lean, it starts to sound like a detraction from creativity and innovation. My goal today is to prove to you that the opposite is true: Lean strategies truly lead to more creativity and more innovative thought, while also cutting waste and saving on the bottom line.

LEAN RULE 1: STRUCTURE EVERY ACTIVITY

"Okay, **the first rule of Lean is to structure every activity**. Now, I know what you're thinking: How can structuring activity *add* to innovation? So, let's take a look at an example." Marcus advanced his PowerPoint to show a slide showing an engineer staring at a drawing.

> Many of you in the room are familiar with this image. You've just completed a drawing for a project you're working on, and it's time to document your progress and push it into the next phase. Does your company have a precise structure in place on how to accomplish this?

Marcus paused and glanced around the room. Some folks were nodding their heads while others looked at one another nervously.

"For those of you looking around or feeling confused, please don't worry. You are not alone in this," Marcus laughed. Many companies like to leave these kinds of steps and processes up to the "creative mind" of their employees. Many times this is a conscious choice of empowerment, not a lazy choice of indifference. Sometimes, this works out just fine; however, most of the time it does not. Unstructured processes like this can prove problematic. Let's go back to our example. So, you have a design in front of

FIGURE 4.1
Confusion

you that you need to document and send to production. Now, let's say that two other people are working on a similar design for production to review. If the three of you document your progress using different templates, symbols, or software, how does production know how to decipher each drawing? In other words, which method is best? Which method ensures that the process is completed correctly and timely? And, most importantly, when some of the methods are performed incorrectly, which one is causing the problem? Which process do you improve?

Marcus advanced the PowerPoint to another slide with a group of people looking confused and worried.

"How many people are familiar with this image?" Marcus laughed as most of the people in the room raised their hands.

That's what I thought. I have consulted with many companies, and one of the biggest complaints I hear from employees is their struggles dealing with sick or recently resigned coworkers. Without structure, the other employees are left to pick up the pieces. Because they relied on their coworkers' knowledge of the process and not of a standardized method, they had to analyze how and why things were being done. This isn't innovation; this is chaos.

FIGURE 4.2
Chaos

Marcus advanced his PowerPoint to the next slide. This time, an image of a disapproving boss appears on the screen.

"Okay, I won't ask how many of you are familiar with this image." The room laughed in response. Here, we are dealing with a boss who just realized that his employee dropped the ball. Now, this may have been prevented if there was a strong structure for moving projects down the timeline in place, but there was not. Now, both the employee and supervisor must come up with a fix for the issue, thus adding to the chaos.

FIGURE 4.3
Get ur done – Now!

The last three slides exemplify the way that process variation causes clogs in a steadily flowing system. Process variation is why you knock it out of the park one day and fall flat on your face the next. It's why some customers, or in this case, production team members, love you and others hate you. Just think if you could consistently hit those home runs and keep all your customers happy – every time! Finding a method that consistently performs at its best will produce high-quality products. This is the only way to ensure that every time that process is used it will be correct, on time, and high-quality, no matter who performs the task.

So, what can be done with the case of the drawing? To be compliant with the Lean rule, you would first **specify the activity's content**. What needs to be done? What is the process? What are the materials needed to complete the process?

Once you know what needs to be done, then you can **specify the order and the timing of the activity**. In what order does this need to be done, and how long should it take? These instructions should be specific and easy for anyone to understand.

Finally, to take it one step further, **there should be a clearly defined outcome**. What results are expected from following these steps? The outcomes should be easy and predictable. If you can clearly answer all of these questions, then you have a well-structured activity. At any given time, an employee can assess what is being done and how it is being done to ensure productivity, and make improvements.

For a team to accomplish this, they need to establish **high agreement** about the best method to accomplish their goals. Establishing high agreement is the Lean principle of standardization. Now, why did I say "current' best method?" These methods should be evaluated for efficiency, quality, and accuracy on a regular basis. When a team member discovers an improvement or better way to perform the task, it should be evaluated by the other team members to be sure it is valid and does not have any negative effects upstream or downstream of the current process. Only then should it be adopted by everyone. This evaluation takes place as part of Rule Four: Improve through Experimentation, which I will discuss in a few minutes.

With this concept in mind, we can see how production would benefit from the engineering team having a standardized documentation and design process. They would be able to easily read each document, decide what needs to be done, and execute their end of the process without incident. In addition, this standardization and structure will allow them to easily identify when something hasn't followed the protocols.

> With this Lean rule, new discoveries of innovative ideas or improvements will benefit the entire team and company, not just one person. This is a powerful process that truly empowers teams. It should be taught, supported, and encouraged by leadership. This is the process used every day by organizations around the world, such as Toyota, to make daily improvements. Without the stabilization and empowerment of rule number one, improvement is haphazard or non-existent.
>
> So, how is this related to innovation? The only thing that prevents chaos is order, and order bolsters innovation. **Chaos prevents us from detecting**

variations between desired and current workflows, which then prevents people from identifying new ways of doing things to be more efficient or productive. See, it's not a matter of "this is the way things are done", but a matter of "if we want to accomplish X, *should* this be the way things are done?" Structure makes things repeatable, which helps people identify any snags in the process. If one is found, folks can brainstorm and collaborate on new ideas. This is how companies grow, and innovation occurs.

Marcus advances the slide to a giant "2" on the screen.

LEAN RULE 2: CLEARLY CONNECT EVERY CUSTOMER–SUPPLIER

The second rule of Lean is to clearly connect every customer and supplier. Every process has a customer and a supplier. Customers can include the end-users of your product or the vendors who purchase your product. Suppliers are often used to acquire items needed for internal production or processes. Most importantly, this also includes internal customers and suppliers.

Every process begins by collecting information or material. In some cases, a process might receive both in order to begin. Once this information or material completes the process, it is then sent onto the next location. For example, if a customer calls to place an order, the process begins as soon as the company's salesperson picks up the phone and collects the information about the order. That information is then entered into an electronic order form of some kind and sent to an order fulfillment specialist. This process might look slightly different should a vendor be calling to place an order, or if that order was being placed internally.

Considering all of the touch points a company might have with customers and suppliers, it is important that each and every one of the internal and external processes has a **clearly defined and easily understood customer-supplier connection**.

Marcus advances his slide to an image of a person who looks frustrated, their phone pressed tightly against their ear as they hold their face in their hand.

"I think all of us are familiar with this image," Marcus continued. We should all strive to be companies who never leave their customers on hold

FIGURE 4.4
Bad connection

for more than a few moments. Further, we should strive to be companies that empower our customers to find the answers they need. **The information that passes between our customers and our companies should be clear, direct, and binary**. Consider implementing "If… then…" statements into your work. For example, *if* they ask about pricing, *then* they will be sent to the pricing page. Any questions they have about pricing should be answered on that web page. The purpose of this exercise is to make things as easy and seamless for the customer so they don't get frustrated with *the process* and end up looking for products elsewhere.

Similar to the last rule, this rule is about eliminating chaos. It requires making sometimes ambiguous and convoluted processes streamlined and obvious.

Marcus advances the PowerPoint once more. A woman is standing by a desk with a phone in each hand and one ringing off the hook behind her.

FIGURE 4.5
Crazy

When we don't have clear connections to our customers and suppliers, we risk losing our manpower. For example, let's say you have a sales manager who is typically responsible for budgets and quality control. If there is an unclear process to connect the customer to the answer they are looking for, you will likely see an uptick in people saying they would "like to speak to the manager". Under these circumstances, the manager would be unlikely to adequately complete his or her daily tasks. This causes increases in stress levels, decreased motivation, and unhappy customers and employees. I don't know about you, but this doesn't sound like an innovative environment to me. Instead, it sounds like one where people need to put out fires and end up working from behind. Having a clearly connected customer-supplier process eliminates the mess, allowing time and space for people to tackle the things that truly matter, like being creative.

Marcus clicks the PowerPoint once more to an image of a child sliding down a bright yellow slide. The audience chuckles at the choice.

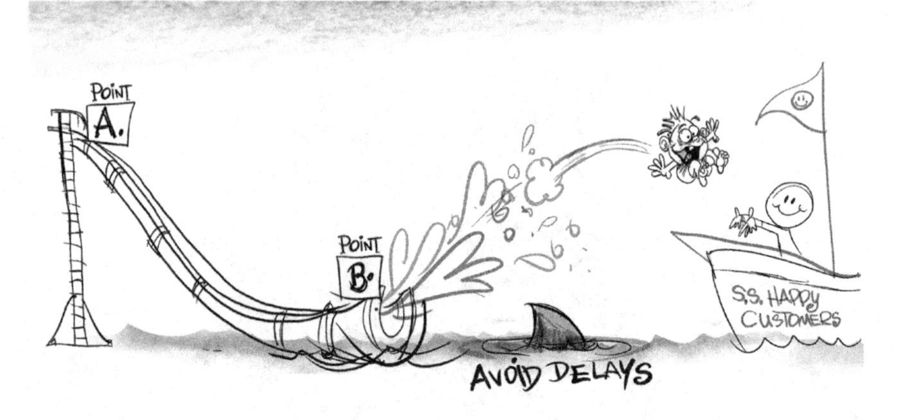

FIGURE 4.6
Avoid delays

LEAN RULE 3: SPECIFY AND SIMPLIFY EVERY FLOW PATH

Okay, so you might be wondering what this represents. Let's just pretend for a moment that this slide was actually shaped like a zig-zag. The goal is still to get to the bottom of the slide on the same incline. To make this happen, the engineers need to make the slope more gradual. Now, when we put that same boy in the slide, and what happens? He still goes down the slide, only much slower. **The third rule of Lean is to specify and simplify every flow path**.

For us to have effective flow paths, we must eliminate all of the unnecessary stops and turns a project or activity might take to get to the end. Whenever possible, all material and information flow throughout your organization should follow a single direct path to get from process to process, or from each supplier to each customer. An A to B flow path is much simpler to follow and is much less likely to produce errors. The goal is to eliminate multiple paths and make the possible available simple and direct.

This is very similar to the first rule in that having simple flow paths eliminates chaos and variability in your process. This kind of simple pathway can help identify the root cause of any potential defects in a process, making them less likely and easier to spot. Innovation is impossible during chaos, and without simple structures ready in place, chaos is waiting to step in and take over.

FIGURE 4.7
Shark

LEAN RULE 4: IMPROVE THROUGH EXPERIMENTATION

For those of you who have been following along, this will make perfect sense. For those of you who have been struggling to keep up, this should bring you up to speed. **The fourth and final rule of Lean is to improve through experimentation**. As you all probably know, experimentation is one of the main tenets of innovation – you can't have one without the other. The same can be said for Lean as well.

The first three rules provide the stability required for experimentation to take place. The scientific method is required when making improvements to activities, flow paths, or connections between customers and suppliers. Each role is a breeding ground for innovative thought and ideas, just not in the grandiose eureka moment way we've been taught to think about innovation. This kind of experimentation is more intentional and backed by the scientific method. Essentially, this final rule of Lean sets the expectation that **everyone and everything can improve**. It **encourages constant movement towards an ideal state where the goal is to provide on-demand goods and services without any defects or waste**.

> Throughout your experience with us today, you will learn more about the principles of Lean and how each of those principles applies directly to creating an innovative work environment. You will learn skills and strategies that you can apply immediately to your company's culture. These skills can take your company to the next level of Lean and innovative thinking. Welcome to the Lean Strategies for Innovation Conference. Thank you.

The audience erupted in applause as Marcus finished his speech. HydroTherm's logo flashed on the screen behind him as the lights turned back on, and the audience dispersed to find their workshop rooms.

"That was fantastic," Andrea praised as Marcus rejoined the group. "You did a great job connecting Lean and innovative practice. I think you did your part to set the tone for this conference!"

"Thank you," Marcus smiled.

"I know I took a ton of notes to take back to my company," Michelle agreed. "Thank you so much for speaking this morning."

"I'm glad you felt like you got a lot out of it," Marcus replied. "Best get going to your workshops. They start in a few minutes."

Mary Lynn, Michelle, Andrea, and Alexander each chose a workshop based on a Lean principle they wanted to work on at their company and agreed to meet up later to exchange notes.

QUESTIONS TO CONSIDER

1. Have you experienced any innovation myths?
2. How do you see Lean fitting into your ability to innovate?
3. What activities could you better structure to increase innovative ideas at your company?
4. How could structuring activities benefit your company?
5. What processes could be simplified at your company?
6. What unnecessary steps could you eliminate from your production or service process?
7. How do you make room for experimentation at your company?
8. What benefits have you noticed from experimenting with processes and procedures?

Part II

Leading Innovation

5

Leader Accountability and Development

"I thought things were going so well," John admitted. He looked even worse than the first time he and Andrea met. His eyes carried dark, swollen bags underneath them, begging for sleep. "I tried implementing some of the changes you recommended, but it seems like my teams are even more frustrated now than they were before."

"Well, I'm glad to hear you're trying! What are some of the changes you implemented?" Andrea asked, sipping her tea carefully.

"The biggest change I made was creating a learning and development team. I tasked them with coming up with fresh ideas and gave them a small budget. They have yet to come up with a profitable idea," John scoffed.

MANAGER BURNOUT AND FRONTLINE INPUT

I'm sorry to hear that it's not going the way you planned."

"How did you decide who would be on the learning and development team?" Andrea asked.

> That was the easy part. I chose a manager from each department to be on the team. I wanted to reward them for their hard work and show them that I believe in their capabilities. I'm just disappointed in their output. It doesn't seem like they are coming up with much,

DOI: 10.4324/9781003318354-7

John shook his head.

Andrea pursed her lips together and debated whether or not she wanted to challenge John on his decision.

"Andrea, I called you for advice. Don't hold back now," John laughed.

"Well, I'm wondering what other responsibilities your managers currently have. Do you know how many of them actually wanted a role on this new team of yours?" Andrea asked.

"What do you mean? Why wouldn't they want a role on this team?" John asked.

I did the same thing when I created my learning and development team. I wanted the best of the best on that crew to ensure that they would develop the most innovative decisions. So, I added all of my top producers to the team and moved forward, full steam ahead. I didn't ask anyone if they wanted to be on the team or offer it as a perk; I simply assigned my managers a new responsibility. I didn't even offer them any benefits. Much like you, I assumed that this was something they wanted and that they would simply benefit from the experience itself. What I didn't account for was burnout.

> **Managers are often carrying the weight of many different responsibilities. They are trying to manage teams, keep up with processes and procedures, communicate across the company, and then take on their own projects. Offering a place on a learning and development team to a manager is one thing, assigning them a new role without consulting with them can have negative impacts on both the new learning and development team as well as the departments they manage,**

Andrea explained.

"I hadn't thought of that," John mulled over Andrea's words in his head. "But if not my managers, then who belongs on a team like this?"

> That's a great question! I recently took one of my employees to a great conference called Lean Strategies for Innovation. **During one of the workshops, they discussed the Lean principle of Directly Observing Work as Activities, Connections, and Flows. Essentially, they recommend gathering information about activities at the point where that activity occurs. This will help you gain a complete understanding of how work is being completed.**

"Okay, how does that relate to my learning and development team?" John asked.

Well, as you can imagine, many managers and owners were a bit put off by this notion of observing every point of activity at their company. That would be incredibly time-consuming. Sometimes, however, it is required that top-level folks observe the work to ensure it is operating correctly. Going back to your example, sometimes it is important for managers to be involved in these decisions. What the managers at the conference weren't thinking about were their frontline workers.

> **The people on the front lines of your company are the most intimately involved with the organization's policies, procedures, customers, products, and services.** These folks are on the front lines every day, directly observing the work. Because of this, **they are best positioned for identifying innovative improvements.** If I were you, I would consider identifying a large group of people that might do well on a learning and development team from all over your company, not just management. Consider holding brief interviews with each person to gauge their interest as well as their workload to set the team up for success,

Andrea finished.

One Big Idea vs. Many Small Ideas

> I see what you're saying. I do feel a little hesitant to put junior people on this particular team, but I guess if some of my more senior folks are involved, it might make for some good decision-making. Hopefully, it will at least be better than what they've already done.

While John seemed marginally less frustrated, he still seemed lost.

"Well, what have they already done?" Andrea asked.

"Their first idea was to put half of the money I gave them into a pot for the rest of the company to invest in their own personal development," John scoffed.

"I don't think I understand. Why is that a bad thing?" Andrea asked, knowing this was one of her recommendations.

"It's not a bad thing by itself," John replied.

> It's just that I was planning to do that myself. I'm frustrated that this was their first decision as a new team. I expected groundbreaking research and ideas to be developed with that money, not to do something I was already planning to do in the first place.

"It sounds like you're expecting a lot from them," Andrea started.

> It doesn't have to be a lot, but it would be nice if they could come up with *one* big idea to pitch. I just want to feel like I'm going to be getting a return on investment from this new team,

John explained. Andrea could tell he was feeling eager for something to happen.

> A few things come to mind here. The first is that you can't rush innovation. You should know that better than most people. It takes a lot of time, hard work, and trial and error to truly work through and develop "big ideas". Second, I would advise you to consider encouraging them to start with small ideas first,

Andrea said.

"Small ideas? What good will that do?" John asked.

"Do you remember Nicolas Rimsey?" Andrea asked.

John's expression looked as if he had just taken a sip of sour milk. He rolled his eyes and made hard eye contact with Andrea.

"What does he have to do with anything?" John asked.

Nicolas was an up and coming mechanical engineer at TechLoPlex where John and Andrea first worked together. Nicolas took an immediate liking to John, and the two quickly became friends. John was an idea man. He was constantly cooking up a new way to approach a project or concept, and Nicolas loved being a part of the action. After a few months of working on a particularly challenging project, John had a huge breakthrough. If he could make it work, it could become the next big idea at the company. At lunch the next day, John shared his excitement with Nicolas who listened intently. Though John didn't give it all away, he gave Nicolas enough information to steal the idea. A few weeks later, Nicolas presented the idea to the CEO as his own.

FIGURE 5.1
Gems of innovation

> Think of Nicolas as any other big company in our sector. Big companies like to steal big ideas. **Big ideas are typically highly visible, and because of their popularity, they can be easily recreated**. Now, imagine your company comes up with a series of smaller ideas. **Small ideas are not as visible, which makes them harder to duplicate elsewhere**. If you start stacking these small ideas on top of one another, suddenly you have something that's much more innovative and protected from idea theft,

Andrea explained.

John's expression told Andrea he wasn't quite convinced.

"I recently read an interview with Art Fry, the inventor of Post-It notes," Andrea said.

> During the interview, he mentioned that the number one reason he thinks his company is still a leader in the sticky note industry is because of the small, frontline ideas developed by the people in his labs and manufacturing plants. **Small ideas matter just as much, if not more than the big ones**.

FIGURE 5.2
Big ideas

Innovative Mindset

"Wow, I didn't know that about Post-It. But Post-It by itself is a big idea. It's okay to stay focused on the smaller ideas when you have one as big as the Post-It," John countered.

"John, it's only been one month since we last met. They need time and resources to start producing on any kind of significant level," Andrea explained. "In the meantime, I'm curious how you are instilling learning and development concepts throughout the company?"

"I don't follow," John admitted. "I have a learning and development team. How else can I distill these concepts into my company?"

"I'm more so wondering how your folks have been instructed to bring ideas to the table at learning and development meetings or with learning and development team members," Andrea corrected.

> The learning and development team is responsible for learning and development. Their sole purpose is to come up with innovative ideas. Everyone else needs to stay focused on the work in front of them, or else everyone will be trying to come up with something new, and the current work won't get done,

John explained, feeling confused.

"Unfortunately, that's a myth that a lot of leaders buy into. In reality, **innovation is everyone's role**. Remember when I said that I wanted to develop a learning culture at my company?" Andrea asked. John nodded.

> I was able to do that by including everyone in the process. I agree that not everyone should be scavenging the internet for the latest in our industry on a daily basis, though it couldn't hurt everyone at the company to be in the know. Instead, I encouraged everyone to come up with ideas and use the think tank meetings as a way to present those ideas to the rest of the company. I made it known that innovation was everyone's responsibility, and I truly expected everyone to contribute in some way. As a leader, it's important to realize that **everyone's voice is important**, not just the people you put on a committee or the managers at the top. **Every employee helps push your company forward, and they should be treated as such**.
>
> It's not that I don't think everyone's voice is important. I do. I guess I was just seeing it as too many cooks in the kitchen. If everyone is spending time in research and development, I worry that there would be a lot of disagreements. I don't want all of my people tied up in the weeds when we need to be producing new products,

John clarified.

High Agreement on What and How

"Ah, I think I have another Lean principle that can help here. It involves **establishing high agreement on what and how**. What kind of expectations and parameters did you provide the team before they got started?" Andrea asked.

"Parameters?" John asked rhetorically. "I'm not sure there are a whole lot of parameters for a learning and development team. I tasked them with researching what's going on within our industry and offering ideas to keep our company at the cutting edge."

"I think that's a great mission for a learning and development team, but they likely need a little more than that to sink their teeth into. Did you give them any expectations or goals to meet?" Andrea asked.

> I suppose I'm not one hundred percent certain what to expect from them. I'd like them to produce big ideas, though I know now that may not be the best expectation to set. Outside of that, I didn't give them any barriers,

John admitted.

> I wouldn't think of expectations and parameters as barriers. Instead, I would consider them as ways to achieve specific goals. **It's vital for you and the learning and development team to come to a mutual agreement on their goals and objectives.** They should feel that the goals are achievable and worthy of pursuing. **After you agree on the goals, you should then agree on the process and tasks that the team will complete to achieve these goals.** These tasks and processes should directly relate to the goals and objectives. I would also encourage you to discuss a potential timeline for achievement with your team. When every person at your company understands that they play a role in innovation and are given a streamlined process for achieving it, your company will start producing at a higher level. **The two Lean principles I mentioned today are designed to create a structure that allows people to think creatively and identify better ways to execute new tasks.** Does that make sense?

Andrea explained.

> That makes a lot of sense. I like that kind of structure, actually. I was concerned that putting too many parameters on them would stifle their individual curiosity and make it seem like I was overbearing, but you're right. I should be setting more rules and giving more direction to keep the projects moving forward,

John agreed.

Leader Input and Control

"Before you start adding too many rules, I'm curious how you participate in the meetings with your team now?" Andrea asked. She realized that John was starting to spin in the wrong direction and wanted to make sure that her correction didn't upset him. She could sense that he was still unhappy with the learning and development team, and she wanted him to feel confident in his decisions.

John leaned back into the booth and pondered Andrea's question. If he were honest with himself, part of the problem was his schedule for the past month. He had been so busy with a new valve design; he hadn't truly been giving the new team the attention they deserved. He didn't even help them set goals. He just told them it was their job to start generating innovative ideas for the rest of the company.

"It's been a busy month," John admitted. "I've been meaning to give them more of my time, but I really haven't been able to. Now that I'm hearing everything you're saying, it's clear I should be much more involved."

> **Helping them develop goals and objectives is an important aspect of being a leader. They will definitely benefit from hearing more specifics about what you'd like to see them accomplish.** Still, it's important for you to give most of the control back to the leaders of that team,

Andrea said.

"Give up control? In what way?" John asked.

> Deciding on goals, objectives, processes, and tasks is the first major step for this team. They will want your involvement to feel like they are on the right track. **Once they are on that track, however, they will want to know that you trust them. The best way to demonstrate that is to let them exercise autonomy over their work.** They need to know that they have the creative freedom to execute those tasks and procedures in a way that makes the most sense for them as a team,

Andrea explained. "Of course, that's outside of your own idea input."

"What do you mean, 'my own idea input'?" John asked with a genuine look of confusion on his face.

"It goes back to what we talked about last time," Andrea recalled.

> This learning and development team is fresh. They are eager to please you, and they want to feel like you're invested in the work their doing, especially if they were assigned to this role. If you bypass their responsibilities as a learning and development body and execute new ideas without running them past this team, it may be demoralizing for them.

"I think I'm still stuck on this one, Andrea," John admitted.

> It's not that I don't trust a team like this with my ideas because I do. I've assigned the brightest minds at my company to this team. However, you know how it is. In this industry, you have to move fast, and sometimes my ideas are the ones we need to be able to move on quickly.

"I think that's an excellent point. How can you **position your learning and development team to move quickly on good ideas that have been presented while maintaining a thorough vetting process for those ideas?**" Andrea asked.

John paused for a few moments to consider Andrea's question. It hadn't occurred to John that a learning and development team would take much more upfront organization and direction for them to operate most effectively. The more Andrea asked him questions, however, the more he was starting to see their value.

> That's a great question. I think I'll set up a meeting with the team to discuss my vision for their output. Once we establish some of those basics, I think I will have a clearer idea on how to position them for success,

John responded after some thought.

> I think that's a great approach. **Making decisions without the input of your team can sometimes result in less productivity**. Providing guidance is key, but letting them carve out the path is going to be an important tradeoff for them,

Andrea responded. "Are you still feeling hesitant about throwing your own ideas into the mix?"

> To be honest, yes, I am. I am starting to understand why it is important to do so, but at the end of the day, I'm the owner. If I have an idea that I want to move forward with, then that's what we are going to do,

John answered bluntly.

> I hear you. Honestly, there are plenty of ideas that my company has implemented that came from my urgency to push it through. The main thing is not to **bypass the systems you put in place**. If you have an idea you'd like to see implemented immediately, I'd still take a few days to let your learning

> and development team take a look. You can let them know upfront that this is something you plan to move forward with, but give them the opportunity to check on the research and offer feedback on how they might improve the final product,

Andrea encouraged.

"Well, I'm always up for ways to improve our products, especially if it makes them more competitive," John replied. "I'm feeling equal parts nervous and excited about this new direction. It's a big change for me, and I'm worried that people won't speak up if something isn't working according to plan."

The After Action Review (AAR)

"Ah well, I have just the idea for you," Andrea laughed. "I knew these Lean concepts would come in handy."

Andrea pulled a small notebook out of her bag and flipped to a page she had titled AARs.

"Okay, I'm assuming you know what an AAR is, right?" Andrea asked.

"After Action Review, yes," John replied.

"Okay, great. Do you use them at all with your company?"

John smiled, "sometimes. They tend to be brief meetings. Typically, I only want to know how it's going and what the plan is moving forward."

"You've got part of the AAR process down," Andrea laughed.

> You want to do this review at the end of a project, not while the project is still in process. During the Lean Strategies for Innovation conference, they discussed AARs as a way to embody the continuous learning mindset we've been talking about. **The AAR process is broken down into four parts. The first step is to remind yourself of the goal for the project team. What were they supposed to accomplish? You may also want to touch on the ideal state, which is the strategic direction of the project team**. What was the deadline? What was the final product supposed to look like? Discussing the future state gives everyone in the room the context they need to make the rest of the AAR most effective. **The second step is to discuss the current state. What did happen?** Why did we miss or make the deadline? How did we make sure the product followed the correct schematics? Addressing the current state of the project next encourages the team to think through both the things that went well and the things that didn't go according to plan. The problem with AARs is that some people stop after this step.

> I think I'm guilty of that at times. I hear what was supposed to happen, and I ask my team to identify any discrepancies from that and why they happened, but I don't really push them to think much further,

John admitted.

> I think that this happens a lot, but it's also a mistake because you could miss out on valuable learning. **Discussing the target state happens next. This means asking what your team learned from this project, as well as what can be sustained and improved upon in the future**. As you can see, this helps you figure out a plan for your next project's future state. **The last step involves coming to a high level of agreement. In this step, you want to ask what needs to happen differently in the future to avoid the identified weaknesses and build on the identified strengths of the team**. Everyone should agree with the changes moving forward to sustain maximum productivity. What I loved about this concept was how the workshop instructor linked it back to Lean and innovation. Lean strategies are tools to build a structured way to experiment and improve continuously. An AAR is just another tool to bolster that kind of innovation over time,

Andrea explained.

> After this workshop, I realized that I had been doing AARs wrong myself. I took this model of asking questions back to my own company, and it's amazing how quickly things seem to be moving in a more productive and streamlined direction.
>
> The more I am listening to you, the more I realize that there is a ton of room for growth at my company. I would love to learn more about what we can be doing to stay competitive in the market,

John commented.

Andrea could see John's excitement but was worried that it was directed only at the things his employees could change and improve upon, not necessarily himself.

"Listen, I have to admit that implementing AARs on this scale was an incredible asset to my company, but it was also an extremely hard pill to swallow at times," Andrea warned.

"What do you mean?" John asked.

> It didn't take long for my team to trust me after I started implementing the AARs. I took them seriously and sometimes coached employees in

> developing new streamlined processes. It was great, and my employees were grateful. After about four months, I received an AAR I will never forget. In the AAR, it stated in black and white that I needed to do a better job of communicating changes down the pipeline. Like many leaders, I had implemented AARs expecting to find ways to improve for my company, but not necessarily for myself. The thought hadn't even crossed my mind,

Andrea admitted. She recalled receiving the AAR vividly and remembered feeling angered by her employee's accusation. "I considered calling him on the spot to demand an explanation. I felt I was owed an explanation."

"Sounds like you were, especially if this person had never come to you directly in the past," John reasoned. "It sounds backhanded to add it to an AAR."

> To some extent, I agree with you. At the time, I was very vocal about my open door-policy for discussing issues at the company. I thought to call him but wanted to sift through my recent correspondence before getting him on the phone. As I read through my emails, I realized that there was some truth to his statements. I had informed the engineering and production team about a change to the product, but not the sales team. Due to time and budget cuts, we decided to create a simplified version of the product. The salesman who wrote the AAR comment was using some of the more complex aspects of the product during his pitches for potential clients. He didn't know about the changes until after he had already told a few clients what they could expect when ordering the product. I made an assumption about communication flow that could have been avoided by simply copying him on the same email I sent to engineering and production,

Andrea finished explaining.

> This AAR completely changed the way I perceived my communication. I took a few ideas to the research and development team and let them turn it into an incredibly intuitive process. I'm telling you all of this because the things you read in the AAR may not always be easy to see. Before you implement something like this, it's important to know yourself well enough to know how you might respond to feedback.

John knew that this was one of his weaknesses. He liked feedback, light touch and actionable across the company. Anything that pointed directly at him always felt like an attack.

Leader Feedback

"What are some strategies you use to prevent yourself from seeing criticism like this as an attack?" John asked.

> The first thing I did was **replace the word criticism with the word feedback**. Even though many people use the phrase "constructive criticism", I think just hearing the word can sometimes cause undue stress. Feedback, on the other hand, doesn't sound quite as harsh or judgmental. The second thing I do is **reframe the feedback**. Even if it's written in an aggressive manner, which is extremely rare, I remind myself that the feedback is there to make the company better. People who provide feedback want to see things improve. I see that as someone who is dedicated to the mission of the company. If they weren't, they wouldn't have taken the time to write it out. I find that the folks who have a lot of feedback are often my most dedicated employees,

Andrea replied.

> I also **separate the feedback from the person leaving the feedback**, especially if you aren't particularly fond of that employee. If you have a tight-knit community, this person has likely already had their feedback validated by others. It's likely they aren't the only person to feel that way. Separating the person from their piece of feedback can give you a bit more perspective.

"What if you don't agree with the feedback?" John asked.

> Since I've implemented this system, **there have only been a handful of times where the feedback was completely invalid or irrelevant**. I started following a few steps after the first AAR. First, I **read the comment a few times to make sure I completely understand what the comment is trying to convey**. If I still feel any confusion over the comment, I request more information. Once I understand what they are trying to communicate, I **review the process**. With the communication feedback, I reviewed my notes and emails to gauge how egregious my communication error was and whether or not it applies to the rest of the company,

Andrea replied.

"What do you do after reviewing?" John asked.

"It depends on how I feel about the feedback. If I agree with the feedback, but don't think it's generalizable to the whole company, I offer to work with that person to devise a way to rectify the issue. For example, if the engineering team feels my editing process is unclear, I'll encourage them to standardize the process and pitch it to me. From there, we strive to meet a high level of agreement by tweaking back and forth until we come up with a plan that works best for everyone.

"If I agree with the feedback and it can be used across the company, similar to the communication issue my employee brought up in the AAR, I will empower that employee to take their ideas to the research and development team. From there, they create a process for everyone to follow to make things work more efficiently.

> Now, if I completely disagree with their feedback, I offer them the opportunity to meet with me one on one to discuss the issue in a safe environment. This allows both the employee and me the opportunity to work through the issue together. If they left the feedback, even if it's wrong, it's still coming from somewhere. As the boss, it's my job to get to the bottom of it,

Andrea answered.

> The important thing to note about my response to feedback is that I **always follow-up**. Your response to feedback, or lack of, is often more important than the feedback itself. Your team needs to feel that they can provide feedback to anyone, even you, and not have it used against them. This is how they develop trust with you and each other. Trust is an important part of good feedback and building a learning organization. They need to feel safe in providing this feedback, and they need to be heard.

"It sounds like you get a lot of feedback even after all these years," John commented.

> Well, that brings us right back to where we were with our last conversation. My company is a learning company. And **because we are always learning, we are always providing feedback to one another. Because of that, we are constantly growing**. That's the beauty of the culture I've built at Alastar. We just keep getting better,

Andrea smiled widely.

"I certainly like the idea of owning a company that is constantly growing," John admitted. "I think this AAR thing could be a great way to start the ball rolling on some of the other issues we discussed today. Do you think we can meet in a few weeks so I can let you know how it's going?"

"Of course! I think you'll find that you can make a lot of improvements right away and that others will take time. Give yourself the headspace you need to implement these changes at a pace that works for you," Andrea offered. "Let me know how it goes!"

QUESTIONS TO CONSIDER

1. Who do you have on your research and development team?
2. How can you prevent responsibility burnout?
3. What do you do to incorporate frontline ideas at your company?
4. How do small ideas play a role in innovation at your company?
5. How often do you assess your research and development teams?
6. How do you know you have high agreement?
7. How often does your company's leadership contribute new ideas?
8. What process do ideas from the company's leadership go through?
9. If different from others, why?
10. What kind of AAR process does your company participate in?
11. How do your employees know their feedback is taken seriously?
12. How does leadership handle feedback?

6

Culture and Values

"How have things been since the Lean Strategies Conference?" Mary Lynn asked in her normal, peppy tone.

Michelle had just walked to the Panera halfway between their companies. It was small for a Panera, but as long as it had the creamy tomato soup, she was happy.

"To be honest, things have gotten very strange at work," Michelle answered, blowing gently on a spoonful of soup.

"How so?" Mary Lynn asked.

> Last week, John pulled together everyone in the company to talk about his vision. It started great. He admitted that he was still learning the best ways to run the company, and he even apologized for not always having the time or headspace to hear us out. His first idea was brilliant. He announced that we would be creating a research and development team to help us remain innovative. He said he wanted the learning and development team to focus on employee learning and the research and development team to focus on coming up with new ideas for the market. A few people scoffed at the similarity between the names. I was just happy that he was finally moving in the right direction. We can figure out names for these things later,

Michelle explained.

"Okay. All of that sounds pretty normal so far," Mary Lynn replied.

> Well, that's where things started getting weird. Essentially, he laid out a multi-faceted vision that included four main parts. First, he explained the thinking behind the learning and development budget. He explained that

DOI: 10.4324/9781003318354-8

> the learning and development budget was created to provide us with more opportunities for learning and skill development. He wants us to pursue professional development activities both inside and outside of work by keeping an eye out for upcoming conferences and workshops. However, he also wants us to identify people at the company we are interested in learning from and ask them to become our mentor.

"Wait," Mary Lynn stopped Michelle. "He wants to create learning and development through mentorship, but he wants you to facilitate those connections?"

> Exactly. I know that's not entirely out of the ordinary. Young professionals are smart to connect with older professionals to learn more skills. Still, he went on to explain that the mentorship program would then provide specialized expertise in a certain subject area as well as ongoing learning and support. I noticed a number of the older members of the company looking around at one another as John described it as a mentorship "program". Clearly, none of them have been approached with such an idea,

Michelle finished explaining.

"Wow. That does sound a bit odd," Mary Lynn jumped in. "Maybe he's still trying to figure out what it will look like? I mean, it is his vision statement. Perhaps he's still in the beginning stages of putting the program together."

> That's what I was thinking, too. So, I disregarded the lack of planning around the concept and kept listening. The second part of his vision involves focusing on smaller ideas while still looking for big ideas. This part is mostly for the research and development team, which he is now planning to shake up a bit since he originally just assigned managers to the team. Now, he says, there will be junior-level folks on the team as well, which he will be announcing in the coming weeks. So, for the time being, the more senior members of the company will remain in the role,

Michelle continued. "The third part of his vision involved the idea that everyone can contribute to the innovative process. All we have to do is check the idea with our supervisors first before we can take it to the R&D team."

Michelle paused to take a bit of her grilled cheese sandwich.

"So, everyone in the company is part of the 'innovative process' as long as their supervisor says it's okay?" Mary Lynn clarified.

"Right," Michelle replied.

> And here's the kicker, he ended the speech by saying that he is implementing an AAR process. He wants everyone to evaluate the processes and procedures in their departments and the company as a whole. He ended his speech by saying that his vision is to have a company of high agreement.

"High agreement with what?" Mary Lynn asked.

> That was the question of the day. As soon as we left, everyone felt even more confused about the direction of the company than we did before the speech. It's tough because it finally seems like he's trying to listen and do something to make the company more effective, but he's trying to do too much at once. My boss is one of the company's senior members assigned to the research and development team and he looks exhausted. We are still short-staffed, so he's trying to carry his workload, part of our junior engineer's workload, and now the research and development team. As soon as the speech was over, I saw him walk right over to John to ask for his time. I can only imagine he's hoping to put one of the junior engineers on the research and development team,

Michelle finished.

"Maybe that could be you?" Mary Lynn asked.

"I doubt it. Although, I would welcome the opportunity," Michelle admitted. "Now, I'm trying to figure out if it's worse now than it was when he didn't realize there was a problem."

> It does sound like he's really trying. A few weeks after the Lean Strategies for Innovation Conference, Andrea sent me to an Innovative Leadership conference in Philadelphia. I learned that innovation is more than just ideas. It's not just having great ideas and saying them out loud; it's all about how you then implement those ideas into your company,

Mary Lynn explained.

"Go on," Michelle said as she took another sip of her soup. "You have my interest."

CULTURE AUDIT

"It sounds like your company would benefit from a culture audit," Mary Lynn explained.

> It's an assessment that helps the leader understand the current state of the culture at their company. Many people don't do this because they think it's more important to know where they are going than to know where they are right now.

"If you don't know where you've been, then you can't know where you're going," Michelle said absentmindedly. "Or something like that."

> Exactly. A culture audit helps the leader identify how and why they have that particular culture, which makes it easier to identify mistakes a leader made in the past. If a leader can't identify the mistakes they made in the past, they are likely to repeat them as they work to develop a new culture,

Mary Lynn continued.

"Which will inevitably result in a return to the original culture," Michelle finished for Mary Lynn. "That makes a lot of sense."

"Right. There are several reasons leaders don't conduct culture audits at their companies," Mary Lynn said.

"Sounds time-consuming," Michelle replied.

"Are you sure you didn't go to this conference too?" Mary Lynn joked.

> The research says that the best way to do this is by meeting with each member of your company personally to best assess how each person views the current culture. **If you want a true embedment of the new culture, you must understand what people like and don't like about the current culture**. For very large companies, this isn't necessarily possible, so they put together a leadership team to go out and assess the company culture. Either way, a culture audit is the first step to embedding the ideal culture at any given company.

"I love that idea. I'm not sure how John would respond to it, though. I'm not sure he's ready to hear about the cultural perception at his company," Michelle responded.

"You never know," Mary Lynn replied. "It sounds like he's trying to make some changes, but he's not clear on his goals. This could at least set him in the right direction."

"What else do you have in your bag of tricks from this conference?" Michelle asked.

INNOVATION EMBEDMENT

> The most interesting concept we discussed was **innovation embedment**. Essentially, the leader of the seminar made it clear that for you **to truly embed innovation into a company, you need to treat it like a core competency of the company**,

Mary Lynn explained.

"What does that look like?" Michelle asked.

> A lot of companies look at innovation implementation in the short-term. What do I need to do to get more innovative results right now? Instead, embedment looks at long-term solutions to sustain innovation over a long period. The focus is on creating a systemic change at the very core of the company that will foster the implementation of more innovative processes and procedures,

Mary Lynn explained.

> This can't be done through trainer-trainee relationships alone. It would be both a waste of resources and the antithesis of innovation to have trainers constantly teaching people how to look for and create new processes and procedures. Instead, innovation embedment seeks to create an environment of self-learning. At companies who actually embed innovation into their DNA, they focus on how each individual can learn and contribute to the whole. Mentorship is a great way to do this because more experienced employees can guide less-experienced employees while still providing the autonomy needed to learn how to do things on their own. The idea here is to foster an environment for new ideas to develop organically. This can't happen if the more experienced folks dictate precisely how things should be done.

"This sounds similar to what we talked about at the Lean Strategies conference. I think it was one of the principles?" Michelle tried to remember.

> Yes, it's **Lean's foundational principle: create a learning organization**. By doing this, companies grow faster and create better ideas than if they weren't focused on learning and development. **Because learning is one of the core competencies, it makes failures seem less fearsome**. Instead of worrying that they might get in trouble, folks from every corner of the company feel like they can contribute to the innovative process,

Mary Lynn said.

> At Whirlpool, they started using something called an **Embedment Wheel** to track how efficiently various areas of the company were driving towards a more innovative mindset. **On the very outside of this wheel was the company's vision and goals**. These two things are the most important. They hold everything together and provide each employee and department the direction they need to move towards innovation. It sounds like this is where John is getting lost.

"I would agree with that. He's trying to define a vision and goals, but he also mentions a lot of smaller things, like money and system alignment," Michelle recalled John's speech.

> See, those things fall *under* the vision and goals of the company. **The pieces of the embedment wheel that Whirlpool followed were things like leader accountability and development, culture and values, resources, knowledge management, change management, recognition, system alignment, and reporting systems**. Whirlpool tracks each of these areas individually to assess how each area is doing in terms of innovation,

Mary Lynn explained.

"That sounds like a lot of work. How did they manage something that big?" Michelle asked.

> They made it a part of their culture and **identified it as a core competency of the company**. They rewarded people who saw opportunities in challenges. They stopped any practice that hindered risk-taking. They stopped punishing people for failures or mistakes. They embraced diversity not only

> in ideas but in the people they hired as well. Every single thing they did, changed, or created was to accomplish their innovation goals,

Mary Lynn said.

"Sounds impressive," Michelle commented. "I assume it went well."

"Well, I assume you've heard of Whirlpool for the reputation it has as a leading appliance company," Mary Lynn joked.

"Indeed, I have," Michelle played along. "I'm going to have to learn more about that embedment wheel. I think John would be interested in having something like that to follow."

Michelle allowed herself to feel excited, but the emotion was fleeting. Her shoulders curled into her chest as her body deflated from her thoughts.

"What's wrong?" Mary Lynn took notice.

"I would love to bring all of this back to John and have him give this a try, but I doubt he would listen to me. Not to mention, there's no recognition for bringing new ideas back to the company," Michelle explained.

INTRINSIC MOTIVATION

"How does John manage rewards and recognition at the company now?" Mary Lynn asked.

> That's just it: he doesn't. Someone could go to him with a great idea, and the rest of the company would never know. It's not just about an email blast or a thank you, either. Sometimes, it would be nice to be recognized by simply feeling more trusted,

Michelle explained.

"So, you're looking for rewards and recognition beyond just extrinsic rewards like money?" Mary Lynn clarified.

"Well, of course, money would be nice," Michelle laughed, "but it's not necessary to make me feel like I'm a contributor."

> I'm sure this won't surprise you, but the conference I went to had a seminar on rewards and recognition. The seminar leader explained that **most people feel more motivated when they are offered intrinsic rewards versus when they are offered extrinsic rewards**,

Mary Lynn said.

"That seems a little counterintuitive, doesn't it?" Michelle asked. "While I can understand that for myself, don't most companies offer raises to reward spectacular contributions?"

> I think many companies do reward folks with money, and I still think that's a viable option, so long as it's not always the only reward offered. **Any company interested in building a brand of innovation should hire and grow folks who are intrinsically motivated.** These folks are high performers because they like the satisfaction that comes with doing something well. They want that feeling of achievement. After a while, money isn't going to cut it for these folks. They want to be challenged and grow. Research suggests that dangling the "raise carrot" in front of an intrinsically motivated employee will eventually turn that employee off. It's called the **Cognitive Evaluation Theory**,

Mary Lynn continued. "**People who like their work and strive for a feeling of achievement typically hang their hats on performance-related rewards.**"

"Similar to that feeling of being trusted?"

"Similar, yes," Mary Lynn responded,

> but even more than that too. An example of a good reward for an intrinsically motivated person would be expanded space to work on creative projects or being invited to senior-level meetings to contribute ideas. Andrea does a great job finding ways to reward folks for their work. For example, she learned early on that I have a strong love for learning. So, the way she recognizes me for my work is to provide more structured learning opportunities around topics that interest me, like this conference. Folks who strive for accomplishment are much more likely to stick around after receiving these kinds of rewards than they would after only receiving a raise.
>
> I would love to have the freedom to work on my own projects. One of John's comments about my work was that I think a little *too* far outside the box, but I always thought that's what made me good at what I do,

Michelle said. "I couldn't agree more with what you said. A raise sounds nice, but having more autonomy or contributing ideas at a higher level really made my heart pump."

> That's exactly what companies who embed innovation into their culture want to hear. They understand that **a team of folks with an appetite for the**

> **work itself is far more valuable than a team of folks motivated by money.** Why do you think so many companies advertise "the opportunity to work on a team with winning ideas" or a team that will "leave a legacy"? They are attracting the intrinsically motivated folks,

Mary Lynn concluded.

"That's a really good point," Michelle agreed.

> One of the reasons I took the job at TrinoTech was because John talked endlessly about his company's lasting legacy in the field. I admired his tenacity and believed that his company could be the next big thing. I wanted to be a part of that.

"Of course you did. Everyone wants to be on the winning team, especially people motivated by the job itself," Mary Lynn agreed. "Not to mention, large-scale rewards often result in creating disincentives."

"Disincentives?" Michelle asked. She couldn't imagine feeling less incentivized by receiving a large reward, even if it wasn't intrinsic.

"Sure. Sometimes, giving out large rewards or recognition creates animosity between team members. This is especially true if more than one team member had a hand in the success of the project that one person is being recognized for," Mary Lynn said.

> Unfortunately, it's common for only one person to be recognized for the success of a project, even if others were involved. Having more intrinsic rewards eliminates the fanfare of a large-scale reward and puts the attention on the success of the work itself.

"Not to mention, eliminating large-scale rewards probably means that companies are more likely to spread out the recognition across the group," Michelle suggested.

> That's an excellent point. Even if one person is at the helm of the project, it's likely more than just that one person played a role in its success. Leadership should find constructive ways to recognize *team effort* in addition to individual effort,

Mary Lynn agreed.

Michelle nodded her head and took the last bite of her apple. If she worked anywhere else, she was confident that she would feel more excited about this conversation, however, she didn't think any of this would go over well with John.

"Am I boring you?" Mary Lynn asked jokingly. "You seemed so into this a few minutes ago."

> I am into this. I'm actually very interested in everything we talked about, but I worry that John won't be. I'd love to take this back to him, but I don't think he values my input since I'm not a senior member of the team,

Michelle replied.

VALUES

"What kinds of things does John value?" Mary Lynn asked.

"First and foremost, John values innovation. He just doesn't know how to implement it," Michelle joked. She took another sip of her soup as she tried to identify a few other things John might value. "I think he values hard work, continuous improvement, and feedback, but only if that feedback doesn't mention anything that's derogatory about him."

"Hold that last thought for a second. First, I want to know how you know these are the things he values," Mary Lynn prompted.

> Well, I know he values hard work because he always seems to gravitate towards the people who visibly put in extra hours. I know he values continuous improvement because he's always talking about innovation and creativity. He is obviously still struggling with how to implement a continuous improvement strategy, but I know it's something he cares about. I also know that he values feedback that isn't directed at him because he almost always asks for feedback about projects, but never about his performance. He wants to know how senior management is doing or how junior engineers are getting acclimated to the company. His values are clear because they are what he talks about most,

Michelle explained.

"Talking about something a lot doesn't necessarily mean he values it. If he starts to put those values into more constructive action, then it would mean he values it," Mary Lynn replied. "**Values in a workplace, especially an innovative one, are typically demonstrated through the constructive *actions* of leadership.**"

"Did your conference cover what kind of values a company should have in order to bolster innovation?" Michelle asked.

As a matter of fact, they did. While there is some variety in the values that successful companies have, it turns out that most innovative companies share seven core values, Mary Lynn paused to take a sip of her water. **The value that these companies focus on most intensely is quality.** Obviously, this makes sense, given the strong push for quality control through Lean strategies. When employees push for more innovative products, their customers and clients receive higher-quality products. Remember, values are demonstrated through constructive action. Quality shows up in how well leaders provide support to teams working on quality improvement, as well as how well they contribute and communicate their own recommendations.

The second value is individuality. Leaders of innovative companies understand that it's the unique perspectives and contributions of individual team members that helps the company move forward. This value shows up in how well a leader leverages his employees' strengths and develops their weaknesses. Often, this is done in the form of professional development opportunities paired with reasonable rewards and recognition for a job well done.

> **The third value is trust.** During the conference, they described this concept as psychological safety. Employees should feel safe to take risks, make suggestions, and reasonably confront the people they work with. A high degree of trust is required for successful AARs, for example. A leader who believes in this value will demonstrate this by taking risks, being highly collaborative on projects that cross his or her desk, and holding people accountable to their workloads. When you have an environment of trust and safety, the process of achieving innovation becomes that much easier.

"I think my company is missing those last two quite a bit," Michelle said. "It seems like he's at least trying with individuality. I can't say the same for trust."

If he's struggling with individuality, then he's likely struggling with the next one too. **The fourth value is creativity.** This isn't the fluffy kind of creativity. Leaders should encourage employees to brainstorm and make good ideas even better by providing adequate space and time for these endeavors.

> **The fifth value is leadership**. This value focuses on how well leaders communicate the company's vision and mission to their employees. These leaders don't look to be the best or seek the spotlight. Instead, they thrive by knowing that their employees are capable of the success they seek for the company. These leaders will try to shine the light on others and remain focused on achieving company goals,

Mary Lynn continued.

> Oh, yea, both of those are big issues. I think you hit the leadership one on the head, though. His biggest problem is that he still wants the spotlight to be on him, and he doesn't do a great job of communicating a strong vision for the company,

Michelle explained. "I think that if he could focus more on the things you were just talking about, the company would be better off."

("It is amazing what you can get done if you do not care who gets the credit." – Harry Truman)

> That's a great point. What you're saying goes hand in hand with **the sixth value: accountability**. Employees in innovative companies take responsibility for their ideas, behaviors, and actions. In the same vein, so should the leaders. This shows up in a leader's decisions. Are the decisions in the best interest of everyone at the company, or just for the leader's bottom line? It's important for both leaders and the employees they serve to know the difference,

Mary Lynn explained.

> **Which brings me to the last value: measurement**. From everything we've discussed, it doesn't sound like John does much with measuring how the company is performing on different projects or how to evaluate the results of certain products on the market.
>
> I would agree with that. As soon as a project ships, John's concern is the next project. He likes to know how the project went and how it's performing, but he doesn't do a lot of measuring to assess the steps that were taken

> to help make the project go well *and* the roadblocks that should be avoided in future projects,

Michele replied.

"That's a big no-no for this last value," Mary Lynn explained.

> Leaders should be tracking not only how projects are going at their completion, but also how they are going as they are being produced. What innovative ideas are going into the creation of the product? What is or isn't working? The idea is to incorporate this learning into future projects. Leaders who demonstrate this value will have formal processes and procedures in place to evaluate how the company is performing at multiple levels, including themselves.

"Again, something we don't have." Michelle shook her head in frustration.

> Well, I guess we do since he started implementing these AARs, but they are so haphazard. Nothing about them is standard, so I'm not sure how he will garner anything useful from them. You have been a huge help today, Mary Lynn. Just the conversation about values alone would be enough of a starting point to help the company move past some of its issues, but I don't know how to get him to consider anything we've talked about.

Michelle looked down at the empty bowl of soup in front of her and wished for a moment she hadn't eaten it so fast. Mary Lynn had so many great things to discuss. She didn't want to cut the conversation short.

"I know lunch is almost over, but I was wondering if you'd be interested in stopping by later to help me practice?" Michelle asked.

"Practice for what?" Mary Lynn asked.

"I'm going to ask for a feedback meeting with John. I think these ideas will help, and I want to see them through," Michelle smiled.

QUESTIONS TO CONSIDER

1. Do you have a clear vision for your company?
2. How do you communicate your vision with your teams?
3. How do you decide on the goals for your company?

4. Have you done a culture audit with your company?
5. What key information do you want to learn from your company's culture audit?
6. How is innovation treated at your company?
7. How do you embed innovation strategies into your company's workflow?
8. What would your company's embedment wheel look like?
9. How do you reward success?
10. How would your company benefit from providing more intrinsic rewards?
11. What are some rewards you could offer your most successful team members?

Part III

Sustaining Innovation and Change

7

Change Management

"Hi John, is everything okay?" Andrea consulted her watch. It was almost nine o'clock on a Thursday night.

"I'm still struggling, Andrea," John admitted. He sounded utterly defeated. "I have been trying to do everything I can to get my company to a better place, but it feels like it's getting worse."

"I'm so sorry to hear that, John. Let me ask you this first. Is anything going well?"

John thought back over the past few weeks. Everything seemed so much more chaotic than it had before. After he announced his new vision to the company, everyone seemed confused and flustered, as if he had told them they were becoming a cereal manufacturer. The only team that seemed to be doing well was learning and development.

"The learning and development initiative is going well," John said quietly. "Several people requested time away from the office all at once, which was a challenge, but it seems like it's making a difference."

"That's good news! I'm glad to hear that people are using it."

"I suppose that's true, Andrea, but everything else is a mess. I feel like I can't manage all of the new initiatives I'm trying to roll out," John said.

"It sounds like you are doing a lot to push your company forward," Andrea praised. "You should be proud of yourself for trying to keep things going as strong as you have been. It sounds a bit like you might be trying to do too much at once, though."

"That's what it feels like," John said. "I only really tried to roll out four things. I just didn't realize how much time they would take up."

DOI: 10.4324/9781003318354-10

> "That's normal, John. I'm not great in this department. I tend to take on too much at once as well, but my mentor, Marcus, has been doing this for years. He is no stranger to managing change at a company like yours. I'm sure he would be more than willing to talk to you,"

Andrea suggested.

"I am interested in doing anything that can help my company," John agreed.

"Great! He's leading a Change Management workshop this Sunday afternoon. I was planning on going, but you should come with me. I'll see if he can meet us for lunch beforehand. Are you interested?" Andrea offered.

"Yes. That would be great!"

"Okay, I'll set it up. In the meantime, try not to worry yourself crazy over this. You're on the right track. We just have to find a better way to narrow your focus," Andrea explained. "Take care of yourself. I'll see you on Sunday."

"Thanks, Andrea. See you then." John hung up the phone and let out a deep sigh. Hopefully, he could get some answers on Sunday.

The end of the week felt like it dragged along. When Sunday arrived, John was dressed and out his door so fast that he ended up at the diner almost 30 minutes early.

"Hi, John." Marcus already had his hand outstretched. "I've heard a lot about you. It's nice to finally put a face to the name."

Andrea patted John on the back as she slid into the booth next to him.

"Andrea tells me that you're struggling a bit with some of the new ideas at your company. Do you want to tell me a little about it?" Marcus asked.

TOO MUCH AT ONCE

> "Sure. So, I've been talking with Andrea a lot lately. At first, I was skeptical about making changes, but I started to realize that my company needed help. I went through all of the ideas that Andrea recommended and came up with a four-tier plan for innovation embedment,"

John began to explain.

> "The first step was to create a learning and development program for everyone at my company. I set aside some money for each person at the company

> to participate in professional development opportunities. The second step included paying more attention to the little ideas. I created a research and development team to help me start implementing smaller innovative ideas while waiting for some of the larger ideas to develop. The third step was to let the company know that everyone is allowed to be involved in the innovative process, even if they aren't directly on the R&D team. Of course, I created a structure where employees would filter their ideas to the R&D team through their supervisors. The last step was implementing an AAR process. This process has been the most difficult. A lot of the feedback has been less than ideal."

"Wow. That is certainly a lot of change you are trying to manage. I can see why you are struggling so much," Marcus said. "I remember having a similar problem with my company."

"Me too," Andrea smiled. "I wanted to do it all at once too."

"We all do," Marcus said. "And we can. The trick is to focus on one thing at a time."

"How do I know what's best to focus on?" John asked.

"I would **start with your vision or mission**. Do you have a clear mission statement?" Marcus asked.

> "Well, my company slogan has always been, 'The Creative Mind and Innovative Spirit Move Us Forward'. I like to remind my folks that innovation is the core of what we do. I just hadn't realized until recently that my company wasn't as innovative as I thought it was,"

John admitted.

"What do you want your vision or mission to focus on?" Marcus asked.

"I think my vision is still worthy of making creativity and innovation a core belief. My vision is to continuously create and provide cutting edge valves for our customers by cultivating an environment that fosters learning, growth, and creativity," John answered honestly.

"That sounds like a beautiful vision," Marcus commented. "Keeping that vision in mind, what would you say is the most important aspect of that four-tiered speech you gave your company?"

John paused. The learning and development fund was running itself. Implementing the second and third steps felt like a daunting task, and perhaps not something that could be implemented easily.

"I think that the AAR process is the most important. I think it will be hard for me to start implementing some of the other ideas without knowing what my people think about the company's direction," John answered.

> "I think that's an excellent place to start. Perhaps start with a company-wide assessment of the culture of innovation. This could give you a great starting point for not only how to format your AAR process but also for future implementation of innovative ideas,"

Marcus advised. "Yes, I think that sounds much more manageable." John took a deep breath. "I feel a lot lighter now. I hadn't realized how much I was trying to juggle."

"It's normal," Marcus commented.

> "We all want to do what's best for our respective companies, and sometimes **it feels like we need to do everything all at once. That's a myth. Pick one thing to focus on, get good at that one thing, and then focus on the next.** It will start to happen faster than you think,"

Marcus said.

"I think it's time for Marcus to get back to the Convention Center for the workshop," Andrea jumped in. "John, are you planning to come this afternoon?"

"I think you could benefit from what we will be talking about. The whole workshop is about helping people manage during times of change at their companies. It seems like you are an ideal candidate." Marcus smiled.

"Sure, I'll be there," John replied.

WILLINGNESS TO CHANGE

I want to begin today's session by talking about the life cycle of an organization. In the start-up phase of a company's development, the growth can be tumultuous. It takes a few months, years maybe, for a company to find its footing. If a company did not plan well or did not have a strong vision, it's likely to go out of business shortly after the start-up phase. Those companies who do their research and work diligently to plan for their future, however, will experience a time of steady growth shortly after the start-up phase has concluded.

FIGURE 7.1
Multitasking

How long the period of steady growth lasts depends on several factors, but eventually companies will notice that they are beginning to plateau. Their numbers will be consistent over the span of a few years, with no signs of growth.

"This, my friends, is the critical moment for any company. When faced with the plateau, how does a company respond? So, I ask you, what is the difference between iPhone and BlackBerry?" Marcus asked.

"Oh, man. I completely forgot about BlackBerry," one of the participants commented. "Exactly," Marcus chuckled. "Do you know what made you forget about it?"

"Yes. My iPhone," the man laughed.

"You are absolutely correct," Marcus replied.

For years, BlackBerry devices were the standard for many top-level executives and A-list celebrities. They seemed to be at the cutting edge of

> smartphone technology. While a bit bulky and without touchscreens, they were still far ahead of their time in terms of handheld cellular devices. Then, change started to occur in the smartphone market. iPhone released their first phone and, while bulkier than the phones we are used to today, it was much more slim and user-friendly than the older BlackBerry models.

Marcus pulled his iPhone out for demonstration and compared it to a BlackBerry photo on the screen.

"The demise of Blackberry is two-fold," Marcus explained.

> Let's start with the simple explanation first: they were unwilling to change. When faced with the plateau, perhaps even a slight decline in sales, they had two choices: stay the course or adapt to the market. Let's take a look at some of these decisions. If you take a look at the Blackberry photos on the screen, what do you notice?

Two photos appeared on the screen. One was labeled a BlackBerry KEY2, and the other was labeled a BlackBerry Evolve.

"Well, for starters, the KEY2 still has a keypad," John offered. "Haven't seen that in a long time. The Evolve looks more standard, like a normal iPhone or Android product."

"The Evolve is still pretty standard-looking. There really isn't much to it," a woman in the back of the room said.

> "I'll help by explaining a bit more about the two phones on the screen. The KEY2 was released in July of 2018. One of its key selling points was a dual-camera system. It was the first time ever that a BlackBerry had a dual-camera system. Does anyone know when the dual-camera system first showed up on the smartphone market?"

"2010?" Andrea offered.

> Close. The first dual-camera system, though elementary, was offered by Samsung in their B710 in 2007! One of the Evolve's key selling points? A virtual keyboard with pre-loaded suggestions and flick typing. Virtual keyboards have been around since 2007 as well, with flick typing and pre-loaded suggestions making their way to the market just a few years later,

Marcus explained. When all of these changes started happening in the smartphone market, BlackBerry made the crippling decision to stand by their old ways of doing things. While they continued to update their software to keep up with the times, their design and features stayed behind. They refused to change, and the rest of the smartphone world marched into the future.

> Today, we barely hear a peep out of BlackBerry. While they are attempting to get back into the market with phones like the one I just described, they are still wildly behind. The performance of their phones continually lags behind their competitors, and they continue to push and peddle a physical keyboard that most of the population would rather do without. When their plateau came, they resisted the changes happening in the world and saw a quick plummet in their sales. Any company that wishes to be truly innovative can take a lesson from Blackberry. If the times are changing around you, it's time to adapt, not resist.

John felt an uncomfortable feeling wash over him. For years, he had been resisting the change that has been inevitable in his industry. Of course, he knew that he would eventually need to make adjustments and change with the tides, but he always asked if "now" was the right time. He had started to feel other companies in his industry pass him by, but he always thought he would be able to keep up. Hearing Marcus talk about BlackBerry, however, made him doubtful.

INNOVATION ADOPTION CURVE

The second issue that BlackBerry faced was a decline in their Late Majority numbers. Late Majority is a term used to label an adoption category on the Innovation Adoption Curve. This curve has five distinct categories, and each one represents a different kind of person who adopts a particular idea or technology. To better understand BlackBerry's demise, we must first understand how the curve works.

In the early stages of a new idea, *Innovators* are the first people to give it a try. Folks in this category tend to have significant financial resources and are willing to put themselves on the front lines. These folks hunger for new ideas and want to be the first consumers of new technology or ideas. Some people see their decision making as impulsive or bold, but *Innovators* derive value from precisely those characteristics. While these folks only make up 2.5% of the adopters for any given innovation, it is incredibly difficult to get an idea off the ground without them.

Once *Innovators* are on board, then come the *Early Adopters*. These people are the trend-setters, just like the popular girls at your local high school. These folks are highly respected members of society, such as famous athletes or actors/actresses. When anyone in this category picks up new innovation, people will undoubtedly notice. Now, at 13.5%, this group is still a small minority of all the adopters across the curve. However, their influence is what goes a long way.

After the *Early Adopters*, we come to the *Early Majority*. These folks spread the idea like wildfire. They are well-integrated into their respective social systems, and they interact with others regularly. While you cannot have sustained buy-in without the *Innovators* and *Early Adopters*, the *Early Majority* are the folks who tip the scale towards massive buy-in. They are a bit more reflective and take more time to deliberate before adopting an idea. At 34%, they make up a large portion of adopters. Therefore, they are needed to place pressure on the next group of adopters.

The *Late Majority* are a more skeptical group than their *Early Majority* counterparts. They are more likely to adopt only if they feel pressure from others to do so. This is where the adoption curve could break down. If the *Early Majority* don't buy in enough to place the pressure on the *Late Majority*, a company could see sales begin to dwindle. This is a susceptible area as the *Late Majority* also makes up 34% of the adoption curve. As you can see, if the adoption curve falls apart here, the product is likely to disappear.

The fifth and final group makes up 16% of the curve and is labeled as the *Laggards*. These folks actively resist change, prefer tradition, and are typically skeptical of the *Innovators* and *Early Adopters*.

As you can see, the critical mass occurs with the transition from Early Majority to Late Majority. This is where BlackBerry began to fail. As other products, such as iPhone and even Android, began to surpass BlackBerry in the market, their *Early Majority* abandoned ship and formulated the *Early Majority* for these newer products. Sure, they already had the *Innovators* who loved the product when it first became available and the *Early Adopters* who saw the value of the product for themselves or their businesses. Still, the product was no longer relevant to the much wider market of the *Early Majority*. With the Early Majority out of the picture, the *Late Majority* never fully developed. Their sales continued to dwindle, and they were unable to keep up with demand,

Marcus finished.

John quietly noted everything that Marcus described. It made sense why some of his valves were beginning to see a drop in market share. They kept making the same valve over and over again, while other companies were taking a similar valve and finding ways to make it better, even if by a small design change. Though John knew that Marcus's program was not aimed at him specifically, he knew that Marcus was right. John was terrified to change too much, and it was impacting his company.

> So, how could BlackBerry have changed the tide for the company? While there are many things they could have done, one would have been to slow down and brainstorm ideas for staying relevant in a quickly changing market. If you attend this afternoon's brainstorming mini-session, we will dive into this idea much more,

Marcus concluded.

BRAINSTORMING

"Welcome to the brainstorming mini-session," Marcus began. He stood at the front of the room with a commanding presence. "Before we begin, what do each of you hope to get out of this session?"

People shouted short phrases like "how to run a brainstorming session," or "when to jump from brainstorm to action." Within a few moments, the room was alight with ideas participants wanted to learn more about. Marcus wrote all of the ideas on the board at the front of the room. As the ideas dwindled, he began combining similar ideas and crossing out suggestions that would not be covered. The board was left with six different topics that Marcus planned to cover.

"These are all great suggestions, and I promise to answer most of them," Marcus said. "Congratulations, you all just participated in a mini-brainstorming session."

A few folks in the room chuckled.

> "Jokes aside, brainstorming is both incredibly simple yet extremely nuanced. As insightful as this mini brainstorming session was for the folks in the room, this would look very different if I were conducting a similar brainstorming

session at my company. **The first question you need to ask yourself before you even begin to brainstorm new ideas is, 'do I have everyone?'"**

Marcus paused for a moment.

"Now, I don't mean 'do I have my whole company?' Obviously, that would be ridiculous. Rather, I **mean, do you have everyone that touches or will touch this idea?** If you are planning on brainstorming ideas for innovation based on your competition, for example, it is valuable to make sure you have everyone in the room to discuss the new ideas. This includes everyone from engineers down to your production and maintenance guys, and everywhere in between. While I can't predict what steps BlackBerry took after they began to see their market share drop, I feel confident that large-scale brainstorming sessions likely didn't take place."

"That seems like a lot of people," John raised. "This is taking a lot of people out of a productive environment to generate a few ideas."

"I'm glad you mentioned that. Many people think they are sacrificing productivity when they have these brainstorming sessions when really they are driving towards a more productive state. Consider this example. You are working on an electrical component for one of your devices. You only include the engineers in the brainstorming phase, and they land on an idea that seems rock solid. They spend the next few weeks drawing it up and working on designs before sending it to production. Production comes back and says that part of the design won't work the way engineering wanted it to, at least not for the cost. Or, perhaps it makes it past production, but sales say the product won't sell at the price point it was designed for. Now, that engineering team has to go back to the drawing board. That's weeks of productivity loss. However, if all of those people were in the room during the brainstorming session, they may have caught some of those discrepancies sooner. Does that make sense?"

Marcus asked.

John nodded. He had never thought of it that way. Looking back, he could think of a few times when something similar to this had happened. Marcus was right; involving more people up front helps bolster productivity significantly.

"Okay, so that's the **first step: gather everyone necessary to the project**. Any questions?" Marcus asked.

No one raised their hands.

"Great. The second step is to structure the brainstorming session. Many brainstorming sessions are based solely on the question at hand, without any other parameters. Here's what I recommend. **First, write the problem you are trying to solve somewhere everyone can see it**. Once you write down the problem, **define the session's goal and identify what's needed to have the discussions that will achieve those goals**. If the goal is to narrow it down to a few solutions, what do you need to know so you can eliminate some ideas? Finally, **set a time limit**. Without time limits, people often go on tangents and derail the conversation from its primary focus,"

Marcus finished.

The next part is the fun part. This is where the free flow of ideas should open for everyone involved in the session. **Create an environment that encourages everyone to share, even if the ideas are a bit unconventional.** The purpose of the session is to **generate an extensive list of ideas and to build off one another's thoughts**. The key to ensuring that the idea-sharing works is to **establish a judgment-free environment**. Folks should be able to readily share ideas, even if they won't work, to keep the creativity flowing. This is similar to the brainstorming session we had at the beginning of this workshop.

Now, If I wanted to continue with the "brainstorming session" I started earlier, I would continue asking questions. Why is it important to know when to jump from brainstorm to action? What is holding you back from running a productive brainstorming session on your own? Leaders like to walk into a brainstorming session with the expectation that they will quickly walk out with a solution. It's not that simple, nor should it be. **The key to creating an environment for your ideal brainstorming session is to ask a lot of questions and refrain from jumping to a solution**,

Marcus explained. When you start a new project or propose a new idea for your company, it's important to **generate effective questions and conversations around the topic. Let the people around you throw out new ideas, and then build off them**.

The last thing to keep in mind is that **brainstorming does not happen the same way for each person**. During brainstorming session, research suggests

> that extroverts speak the most while introverts prefer to absorb ideas, think through them, and make suggestions later. When I first learned this, I started keeping all of my brainstorming sessions open at the end. I **allowed folks in my company to submit any ideas they came up for up to a week after the meeting**. It was drastic how many additional ideas I received after a brainstorming session had ended, and the projects grew even faster. Now, imagine that BlackBerry constructed brainstorming sessions with structured parameters, a strongly defined problem, and additional time to submit potential ideas to address their fall in the market. The flow of ideas would likely have been endless and, perhaps, they would have been able to add dual cameras and virtual keyboards almost a decade earlier to keep up with demand. Okay, I just threw a lot out at all of you. Does anyone have any questions?

Marcus finished.

"Yes. I was wondering if all brainstorming sessions are the same or if there are different ways to structure them," a man in the front row asked.

"That's a great question," Marcus replied.

There are several different ways to set up a brainstorming session. I'll explain the ones that I use most frequently, but I highly recommend looking them up and deciding what is best for your company. The first is called the **Five Whys**. This is a Lean exercise that I picked up as I was learning Lean processes. It is normally used as part of a problem-solving process such as the A3-based problem-solving methodology. It's an exercise used to drill down to the root cause of a problem. You **start with the problem you're having, and you ask your team, "why is this happening?" Once they identify the *why* you ask the same question about the cause of that problem**. Now, the brainstorming exercise is called the Five Whys, but there is no magic to the number five. That is just the average number of "whys" typically asked during this session. You may need to ask "why" three times or seven times. The idea is to ask "why is this happening?" enough times to identify the root cause of the problem. You will know that you are at the root cause when you can answer the question, "If you fix this item, will the original problem go away forever?" From there, you can brainstorm ideas to fix that problem. Used in the correct situations, Five Whys is a very effective problem resolution aid. In the right situations, this is a fantastic brainstorming tool. I recommend checking it out if it piques your interest.

A second brainstorming method I like to use when I'm trying to determine if a new project is worth taking on is called **SWOT analysis. It**

stands for strengths, weaknesses, opportunities, and threats. While this exercise can be done individually, I find that it works great when an entire team of folks comes together to brainstorm each area of the SWOT box. What I like about this exercise is that you don't have to go in order, especially if you know there is one area of the SWOT box your team needs to stay focused on.

The last method I use frequently is **reverse brainstorming**. To be honest, this one is my favorite. It gets the creative juices of my team flowing. I sometimes find that it can be more effective than other more traditional versions of brainstorming. Many times when starting a new project, people come to the table with their doubts first. It's a phenomenon known as the **negativity bias, where our brains are naturally more hardwired to see threats and roadblocks**. I like to leverage this bias. Instead of asking my team to come up with new ideas to solve a problem, **I ask them to come up with ways to make a problem worse or to make sure that a project will fail**. I ask them to generate a list of terrible ideas for making the situation as bad as they can imagine, then ask them to turn each of those bad ideas around to improve the process. For example, if they suggest skipping product testing as a way to minimize product effectiveness, then we've accomplished two things. First, we've identified that testing is an important part of the process, and we now have a place to focus on building in a resolution based on testing. This strategy is a secret brainstorming technique used by billionaires to spark their creativity and get past thought barriers they might be experiencing.

As a bonus, you could also try the **Most Ideas Contest**. It's another secret billionaire technique that **places value on the quantity of ideas over the quality of the ideas**. At first, anyway. Split your team into two groups and give them a set amount of time, say fifteen minutes, to come up with as many ideas as possible. In the end, you'll probably notice that **the last few ideas that each team came up with were the strongest**. They also tend to be the most creative and effective, which helps improve the quality of the generated ideas.

> There are many brainstorming exercises that exist in the world. At the beginning of any new venture, I recommend at least one brainstorming session with your folks to ensure that everyone involved in the process feels heard and on the same page. Please be sure to choose brainstorming exercises that will work best for your company's culture and employees,

Marcus finished.

He took a deep breath and paused at the front of the room, waiting for any additional comments or questions. A woman sitting in the middle of the room raised her hand to speak.

> Thank you so much for your brainstorming ideas. I know there are some great suggestions for me to take back to my company. I have two lingering questions about this concept. The first is how to determine which ideas are worthy of holding a brainstorming session around. The second is how to determine when a brainstormed idea is ready for implementation.

IDEA PROGRAMS AND IMPLEMENTATION

"That slides nicely into the next section we are going to be talking about today. Does anyone in this room have an **idea program** set up at their company currently?" Marcus asked.

Two people raised their hands timidly.

"Would either of you mind sharing with the rest of us how your program works?" Marcus asked.

A younger man with a pinstriped suit stood quickly and faced the room.

> Sure. Hello, everyone. My name is Jaden, and I work for CorpEDU in Ohio. Our company has a unique idea program that we started a few months ago. It took a while to catch steam, but it's been going well since. Our head of HR created a Google form where we can submit ideas for improvement. It allows us to indicate if it's a company-wide, project, or departmental idea. Depending on what is chosen, it is then submitted to someone in that area's research and development team for review. Sometimes, they will follow up with questions, but they do always bring you into the room when the research and development team discusses it,

Jaden explained.

"That sounds like a great system," Marcus praised.

> I would only take what Jaden's company is doing perhaps one step further. A key piece of **an effective idea program is that it should require the idea giver to provide a strong case for their idea**. I see a lot of idea programs

go wrong because they simply ask for the idea, sort of like an idea box. Someone can write "condensed scheduling" on a piece of paper and submit it with no further explanation. This can add to a company's waste, both in materials and time. What I recommend is **designing a structured mechanism for your idea programs.** In Jaden's case, **they created an online form that was designed around his company's needs, and it was easily accessible to all employees.** It's important to design this program to fit the needs of your employees and the company culture. Create the forms, committees, and metrics with your company in mind. It's not important how another company did it if you already know it won't work that way for yours.

Another thing Jaden's company did well was to ask employees to indicate which part of the company will be impacted by the idea. This involves **developing a structure around the idea program**. Employees need to know who the idea will be going to and any guidelines for the submitted ideas. Will there be a budget? What is the approval process? If selected, what would the employee engagement look like? These are all significant considerations to make when developing an idea program. As I previously mentioned, a space to provide an explanation is helpful. Condensed scheduling is a great idea that bolsters productivity when it's done right. If I were reviewing this idea, I would want to know why this person thinks it will work, how it will impact the company, and the steps they believe the company should take to roll out the idea.

"How does this differ from a research and development team?" John asked.
"I like where your head is at," Marcus smiled.

Research and development teams should not be the only people in the company contributing new ideas. **The people working on the front lines of your company have first-hand knowledge and ideas for how to improve workflow**. They are the "experts", if you will, on their jobs. So, they should have just as much of a say as the research and development teams. Now, similar to Jaden's company, it's okay for the ideas to get filtered to research and development. The caveat is that the person who submitted idea should have an opportunity to participate in the process.

Now that we have a better idea of what an idea program is, let's take a look at how to implement this kind of program. A big mistake with these programs is that companies tend to announce that they are happening without providing context or guidance. They send out an email with a brief set of instructions that explains how to submit the idea form. That's it. For those of you with idea programs, what do you think is missing from that practice?

Marcus asked.

A woman raised her hand, and Marcus gestured for her to explain.

> One of the things I think my company did well was to **explain why they wanted to implement the idea program in the first place**. I've worked at companies in the past who would start new initiatives without ever explaining the context. I liked that my company came out and told us that they wanted to see fresh, new ideas from some of our less senior folks. They wanted to see change and to start connecting to a younger market. Not only did it help us understand the "why" behind the program, but it helped us better understand what kinds of ideas they were looking for moving forward,

the woman said.

"That is excellent!" Marcus said.

> It sounds like your company understands that the next step in implementing a program like this is **to educate employees about the program. Employees should know why the program exists and how to use the program as well**. It's important to outline the roles and responsibilities of the people in charge of the program. **Supply step-by-step instructions on how to use the program, and outline any parameters you'd like your folks to stay within**. They should also know how the timing of the program works, which we will get to in a second. Educating employees on the process will significantly increase the chances that your employees will use the program in the first place.

"Anyone care to guess what happens next?" Marcus asked.

"Follow-up," Andrea answered quickly.

"Exactly. You've learned well!" Marcus laughed.

> A lot of people in leadership positions think that follow-up simply means responding with "yes, your idea is being considered" or "sorry, this won't work", but it's so much more than that. Andrea, I know you have a unique way of doing this. Would you like to share it with the rest of the group?

"Sure." Andrea stood to address the rest of the folks in the room.

> At my company, we also have an online form, though folks can submit ideas directly to our research and development team. I have two educational

coordinators on the research and development team who are senior-level managers. While they work closely with the research and development team, their primary responsibility is coaching. Anytime an idea comes in that is off to a good start, but perhaps isn't quite where it needs to be to make it to the R&D team, it gets passed to one of my educational coordinators. They set up a meeting with the employee and coach them through their thought process. Essentially, they help the employee make their idea stronger so that they can resubmit.

Everyone should take note of Andrea's strategy. **Idea submissions are an opportunity for learning and growth for your employees**. Each idea should not only be recognized and shown the appreciation it deserves, but if the idea is worthy, it should be treated as such. Andrea, what is the success rate of your educational coordinators?

"They have about a ninety-two percent success rate of helping employees bring their ideas to the implementation stage," Andrea answered coolly.

"Ninety-two percent," Marcus paused. "That's a robust success rate." Employees who feel their ideas are heard, appreciated, and valued turn into strong contributors to the company's future success. The **continuous follow-up** pieces of the implementation plan cannot be overstated enough.

The last piece of the continuous improvement plan is reflection. Ask what's working well? What isn't working? What can be improved? I use these questions to not only address the idea we are discussing but the idea program itself. **An idea program is only as good as its implementation strategy**. Both the success and failure of the idea program can impact the development of its implementation strategy. Successful ideas should be thoroughly analyzed to identify ways to maximize their success. Failures should be assessed to determine areas for improvement. Both are equally important and should thus have equal attention,

Marcus explained.

John shifted uncomfortably in his seat and raised his hand.

"Every time I hear ideas like this, I get excited," John said, "but then I realize how much more involved they are than I originally thought. How do you maintain an idea program or a research and development team for that matter?"

LEARNING SYSTEMS

> I love this question, and I get it every time I give this presentation. I love it because there is a simple answer that I like to back with some details. The simple answer is this: **you maintain your idea programs, research and development teams, and other innovative projects the same way – by establishing and supporting a learning system**. A learning system includes processes, procedures, and projects that all contribute to the constant learning and growth of both the company and individuals at that company,

Marcus explained.

John glanced over at Andrea, who was smiling. This is exactly what Andrea was talking about the last time the two of them met. It's not about throwing in a research and development team and calling yourself innovative; it's about creating an entire culture that craves learning and growth.

> Now, that's the easy version. The hard part is establishing and continuously supporting a learning system within your company. The first thing to keep in mind is that learning systems encourage learning and development at both the company and individual level. AARs, for example, are great tools to assess employee performance, as well as the performance of your teams or company. It's smaller ideas like this that most people start with when they think of maintaining their learning and development team. It's my stance that we should start broader than that. **In order to truly maintain the system, you need to understand your company's goals and mission**. It's important to talk to key stakeholders in various areas of the company to first establish the needs of the company and any gaps that may exist between company skills and company needs

Marcus began. What is the company's six-month plan? Twelve-month plan? What are the company's primary goals this fiscal year? How far off is the company from meeting these goals, and what does the company need to achieve them? This is an exercise in working from what folks in the coaching world call a "bird's eye view." You need to be able to see your company and everything going on within in from the top down.

> **Once you've established the needs of the company, it's time to talk with leaders of various teams across your company. The goal is to assess the**

kind of learning and training needed to meet the goals your key stakeholders outlined. The important thing to remember is that you aren't trying to assess individual staff members needs quite yet. This is still part of the initial planning phase of the learning system, so staying at the bigger picture is crucial. At this stage, all you are looking to do is assess the feasibility of the company goals based on each team's plans and needs for achieving those goals.

The next piece of this goes back to some of the things we have previously discussed in this seminar. The next step is to **bridge the gap between the goals of the company and the company's current abilities to meet those goals**. Here is where you could employ a brainstorming session with your research and development team, or perhaps ask folks to participate in an idea program to come up with a few solutions. It's important to consider all of your options when generating these ideas. **Many leaders make the mistake of thinking the only resources available to them are costly**. Sure, professional development conferences or workshops are valuable for employees and the company by extension but are certainly not the only option. I recommend creating a list of the various solutions or programs your company could consider as it aims to create a learning system. Consider resources like books, podcasts, or websites. Sometimes videos like Ted Talks or even social media followings can be helpful ways to provide no-cost resources to the learning and development of company employees. It is also vitally important to make sure that you are not simply offering resources without a way to measure their success. It is crucial that you know the goals and solutions you end up proposing, and that you have a way to check back on the progress being made regularly.

What I love about this seminar is that, up until this point, all of this sounds pretty straightforward and relatively easy to implement. The next part, however, gets a bit muddier. It asserts that you **plan for things that can't necessarily be foreseen**. It is for this reason that I feel very strongly about Lean strategies for innovation. There will come a time when, regardless of all of your planning, something will still go wrong. The idea isn't to panic about the mistake, but to have something in place because you've already considered it as an option. **One of the Lean principles requires a systematic-approach to problem-solving that helps identify the root cause of the problems**, rather than putting out fires. What I like about this strategy is that it pulls from the A3 methodology of systematic problem-solving and encourages people to ask "why" questions. Similar to the 5Why brainstorming activity we discussed earlier, this helps companies find the root cause of the problem. However, the answers to these "why" questions are not always obvious.

> For example, if the heater in a building stops working, you would start by asking, "why did it stop working?" After inspecting, you notice that the filter is dirty, but you also notice that the fan seems to be sticking. It's likely that only one of these caused the fan to stop working, but it's impossible to know which without asking another why question. Now you have two possible problems to investigate further, until you identify the root cause. While it isn't ideal to find more than one problem in any system, it is important to create a list of all possible problems and then generate solutions for each. This way, when one of those problems arises, they will have a ready-to-go solution in place to keep the workflow moving effortlessly. Even when things are running smoothly at my company, I like to take use a similar strategy by employing "what if" questions. What if the product doesn't sell? What if the project goes over budget? What if the timeline changes? These questions help create contingency plans so that we can be more proactive should a problem arise unexpectedly,

Marcus continued to explain. Having a structured problem solving methodology firmly in place and the discipline to problem solving (as opposed to fire fight) is the best way to handle the unexpected problems that pop up.

I want to reinforce that it is **incredibly important to identify ways to assess your learning system's implementation**. Before you can assess it, **you need to define for yourself and your company what a successful learning system will look like**. As you've seen already, there can be quite a few moving parts to learning programs, and it will be impossible to go at this alone. Consider the people you have working for you and the responsibilities this program will require. Who are the people who could successfully fulfill those responsibilities? Do they have the bandwidth to help? The team you build around this system is just as important as the system itself. They should understand its vision, be excited to buy in, and eager to offer suggestions.

> This kind of process tends to feel overwhelming. This is not the goal, nor should it be. As I mentioned earlier, **it's important to prioritize tasks and focus on one thing at a time**. I recommend **putting together a timeline**

and prioritizing tasks against the timeline. Create a timeline that is both ambitious and realistic. To maintain a sense of forward momentum, you should feel like you're accomplishing a little bit every day or week, but it should not be so overwhelming that it takes you away from your daily duties. **The idea is to focus on creating short-term goals that build on one another over a longer period**. This bolsters motivation, drive, and significantly increases your chances of a successful implementation plan. I recommend creating quarterly progress reports to send out to the company to reinforce the idea that the whole company is part of one learning system. Discuss challenges and highlight the successes to maintain transparency. Some of the folks in low-visibility areas of the company may not see any of the progress for months. Showing them what you're doing earlier and allowing them to get excited about the process will go a long way for their morale. Are there any questions?

Marcus stepped back to the podium to take a sip of water while he waited for folks to finish digesting the information.

> So, it sounds like there are a few different layers to a strong learning system. One is at a more individual level. Another is at the team or department level. Yet another is at the company level. The last, if I'm not mistaken, is at the results level, or the output level. Is that correct?

John asked.

"I would say that's a strong assessment, yes," Marcus replied. "Why do you ask?"

"I'm wondering what kind of tools you recommend for tracking and assessing progress at each level?" John asked.

MEASURES

That's a really difficult question to answer, considering that every company is unique. There are countless assessment tools out there to choose from. I think it's important to discuss the kinds of assessments you will want to use for your company. I'll go ahead and break down each level and then give an example or two of the measurement tools I use for each of those levels.

Let's start with the output level you were describing. Here, you will want to assess how well your idea management system is working. Are ideas going from idea to inception without any hiccups in the pipeline? If ideas are getting stuck, is there consistency with what stage in the process this is happening? For this process, I often rely heavily on the AAR. Let your folks answer questions about what's working well, what isn't, and what their proposals are for making it better.

You'll also want to assess revenue potential vs. revenue generation at the output level. Is there a gap between the two? If so, what is causing there to be a gap? For this kind of assessment, I have used key performance indicators for the product itself. After a product reaches my sales team, I have them assess their own KPIs against the expectations of the production or engineering team. Once they develop their KPIs, they will then track how the product does against those indicators. They take detailed notes at every step in the process to determine if the product is meeting the performance indicators assigned to it.

At the individual level, I have two go-to assessments. The first is attached to their performance review, and it involves SMART goals. SMART goals are Specific, Measurable, Attainable, Relevant, and Time-specific. I expect each employee to identify at least two SMART goals to achieve in the coming year. They are expected to have monthly meetings with their supervisors to discuss goal progress and course-correct when necessary. Another great way to assess your employees is through 360 Feedback loops. I tend to use this one sparingly as it is more involved, but I value it for its thorough nature. Essentially, each person on a given team evaluates every other person on that team from top to bottom. This way, your peers, supervisors, and direct reports can provide feedback for personal development.

The team level is a bit more difficult as it depends on the team. For example, with my production teams, I have them practicing Total Productive Maintenance (TPM). This is a way to empower my production guys and ensure that all of our equipment runs smoothly at all times. Through this process, they create production sheets, maintenance logs, modeling programs, etc. to assess the needs of the equipment and keep them running at full capacity as often as possible. Because they are constantly conducting a variety of analyses on the machines, and practicing predictive maintenance, we can catch many errors in our procedures before they happen.

The company level is the hardest to get an assessment together for because it is often measured by many different metrics. That being said, I tend to use SMART goals here, as well. As a company, each department comes to me with the goals that they have for the year, which helps me figure out what the company's goals will be in terms of production, revenue, and expansion. I provide quarterly updates on these goals and schedule regular check-ins with key stakeholders to make sure that we are all contributing.

> As I mentioned earlier, these are not the only measures you can use to assess your company's progress. **Each company is different and should be treated as such**. It is best to go back to each company and assess what is best based on the needs of your teams and individuals before making any drastic changes to the way you measure and assess success at your company. Are there any questions on measures?

Marcus finished.

The folks in the room shook their heads.

> Alright, then. We covered a lot in this workshop today, but the key is to start with one thing at a time. It's easy to get excited about these workshops and conferences and try to implement the ideas right away. Choose one thing you want to focus on and start there. I promise you will thank yourself later. Thank you all for coming to the workshop today. Please reach out in the future if you have any questions!

The room erupted into applause as John and Andrea gathered their things.

"How are you feeling now?" Andrea asked.

"Great. I think I know just what I need to do," John answered.

QUESTIONS TO CONSIDER

1. How do you know when you're trying to do too much at once?
2. How can your vision or mission help with idea implementation?
3. How willing is your company to change?
4. How can you ensure the investment of the early adopters and the early majority?

5. How do you use brainstorming at your company?
6. How can you create a safe environment for a brainstorming session?
7. What kind of brainstorming would be most effective for your company?
8. What kind of idea program does your company currently have?
9. Could your idea program benefit from an overhaul?
10. How well does your company communicate new ideas?
11. How well does your company educate employees on the idea submission process?
12. How does your company nurture new ideas?
13. What can your company do to better support employees with innovative ideas?
14. How well does your company assess the needs of the company as a whole?
15. Do you have a learning system?
16. What could you do to better nurture an environment of learning at your company?
17. How do you measure success?

8

Ideas and Talent

The seminar left John feeling motivated and invigorated in a way he hadn't remembered feeling in years. John spent the next week scouring his notes for the ideas he wanted to implement. When it was all said and done, he counted fifteen ideas he was trying to implement at once.

"Now, the real work begins," John muttered to himself.

Next, he compared the notes from his ideas to the mission and vision he had created for his company. On the side of his computer screen hung the following questions:

- **What are your company goals?**
- **Who will help you achieve them?**
- **How will you get there?**
- **What resources do you need?**

John decided to use these questions to prioritize which ideas were best for the company right now and which should wait for more input from his team. The key, he reminded himself, was to slow down and take a more intentional approach to each process. Instead of forcing new ideas through the pipeline and frustrating everyone as he did, he decided to focus on a very small number of ideas that would help pave the way for more idea implementation later.

By focusing on his company's mission and vision first, he was able to come up with a tangible list of goals he wanted to accomplish. Each goal would require a special team of people whose expertise would ensure that goal's completion. John jotted a few names down next to each goal, highlighting the people he felt would be best suited for the next step: small idea implementation.

DOI: 10.4324/9781003318354-11

LEVERAGING TALENT

Michelle's hands shook slightly as she tapped on the door to John's office. She has been working for the company for almost a year at this point and had never been summoned to a meeting with him. The only thing the email said was that he wanted to see her before she left for the day.

"Come in." John's voice came abruptly through the heavy, wooden door.

Michelle entered the room to see John and her direct supervisor, Luca, sitting across from one another in John's office.

"Take a seat, Michelle," John said stoically. "Would you like anything to drink?"

"No, thank you," Michelle responded politely. "I'm sorry, can I ask what this is about?"

"Oh, yes. Well, as you know, I've been attempting to implement a few new ideas here at the company," John said. Luca cleared his throat.

> Okay, I've been attempting to implement *a lot* of new ideas here at the company. Anyway, I've spent the past few weeks talking to the leadership of our teams to find out who their top performers are and what kind of ideas they bring to the table. Luca, here, tells me that you're full of new ideas. Is that true?

Michelle looked to Luca and back to John.

"Well, I don't know if I'd say 'full' of them, but I try to offer new ideas if I think they are worth exploring," Michelle answered confidently. After all of the conversations she'd had with Mary Lynn recently, she didn't want to sound weak now.

"That's great," John remarked. "Luca says that most of your ideas are pretty solid, too. The biggest hurdle is me." John paused for dramatic effect.

Michelle remained silent, unsure what kind of game John and Luca were playing.

> Listen, Michelle, I know you have a lot to offer. Luca tells me that you came back from that Lean conference full of ideas. **I want to apologize for not taking the time to meet with you to discuss your ideas. It seems like you**

> **have a real talent for seeing problems as opportunities for improvement. That's an important quality to have,**

John explained.

"Thank you, sir," Michelle responded. "I prefer to be proactive instead of reactive."

> Now, that's a talent that I lack. I don't think it's any secret that I've been much more reactive than proactive as of late, and I want to do everything I can to turn that around, starting with leveraging our talent. As you know, Luca is a member of our research and development team. He's there because I recognize his talent for being critical of ideas and asking thoughtful questions. Jameel is leading the learning and development team because I recognize his talent in helping others achieve their goals,

John explained.

> Which brings me to you. I've been hearing a lot about your contributions. Luca has been reinforcing these rumors by informing me of your successes. It sounds like you have a talent for creating effective and efficient systems and processes. One of my goals this year is to start streamlining processes across the company by implementing a Knowledge Management System. I recognize that my talents do not lend themselves to this kind of work, and was wondering how you might feel about heading up a Knowledge Management committee?

Michelle sat back in her chair, stunned. This was not the conversation she expected.

> Before you answer, I spoke with Luca about your workload, and he's willing to work with you to find a balance so you don't feel overloaded. I understand if you want time to think about it, but please know that Luca and I both think you are the best fit for this role,

John finished.

"Well, I suppose I'd like to sit down with you to discuss the ideas you have for this kind of system. Once we do that, however, I would love to take on this initiative," Michelle responded.

Finally, her work was being recognized.

KNOWLEDGE MANAGEMENT

Michelle spent the first two weeks of her assignment building a strong team. She started by creating a description of the roles and responsibilities for the committee. She wanted to make sure that the roles were clear, well defined, and achievable. Then, she made a list of the people she had networked with when she first started with the company. She went back through her notes and identified the people who had knowledge or expertise in building knowledge management systems. This process eliminated waste and time by weeding out anyone who might not be interested in the role up front.

Of the folks that she identified, four of them were both excited about the initiative and available to join the committee. At the first committee meeting, Michelle came prepared with an agenda and a general timeline for the next steps.

"I think we should start by interviewing the people at the company who are in charge of knowledge management for their individual departments," Michelle offered.

"I agree," Toby from accounting said. "I think we need to ask what kind of knowledge each department is capturing and tracking."

"I think we should divide and conquer," Michelle suggested. "We can each set up a meeting with two different departments to learn more about how they capture learning. Then, we can compile notes and start researching systems."

"What about an online form?" Clara from production asked.

> I don't mind meeting with people in person, but is there a way we can expedite the process by creating an online form for them to fill out? I'm worried that by only talking to one person in each department, we may miss key pieces of information that other folks in that department could contribute.

The rest of the committee turned to look at Michelle for her reaction.

"I think that's a great idea," Michelle said.

During their next two meetings, the committee put together an extensive form for the teams to fill out. It asked questions about how each department tracked productivity, task assignments, completed tasks, tasks in the queue, and communication. It covered timelines and goals for projects. At TrinoTech, time is a valuable resource, and it was important to

Michelle and her committee that the survey took a minimal amount of time. The committee edited the document down so that it would only take each person no more than ten minutes to complete while still collecting the most important information. That afternoon, Michelle sent the survey to John for him to review. She also felt that it was important for the survey to come from John as the leader of the company.

John largely agreed with the committee's assessments. He loved the idea of online forms for boosted efficiency and felt strongly that everyone should contribute. He encouraged Michelle to copy him on the email to the company to emphasize the importance of completing the form.

> Please keep budget in mind when you start getting some of these surveys back. I anticipate that everyone will likely be looking for different things, and finding software to accommodate all of it could be expensive. If you have any questions about this as you move forward through the process, please let me know,

John said. "Good work."

By the two-week deadline, the committee had received 89% of the surveys back. All of the survey results were compiled into a final document that was easy to review and analyze.

"It looks like a lot of people require timeline tracking as part of their knowledge capture," Clara commented.

"And task delegation seems to be a big deal too," Toby mentioned.

Every need that was mentioned by more than 50% of the company was written on the whiteboard in the room. From there, the committee **prioritized needs by both demand and usefulness across every department in the company**. In the end, they came up with ten key needs. They also had a list of ten knowledge management systems each of these departments was already using.

"Now, for the fun part," Michelle joked. The team agreed that this was where they would divide and conquer. Three of the committee members divided the list of current knowledge management systems between them. They set a common goal for their review: to extensively analyze the benefits and drawbacks of each management system already in use. The other two committee members were tasked with coming back to the committee with an additional two options for review.

They decided they needed to analyze each system consistently. The next meeting's discussion would likely be frustrating and unproductive if each person walked away from the meeting without knowing what metrics each system would be measured against. Together, they decided to analyze each system for the key aspects they identified, plus a section for each program's pros and cons.

The committee created the following grid for their analyses:

Knowledge Management System	
Website	
	Brief Description
Pricing	
Graphics	
Task-Delegation	
Timeline Tracking	
Progress Tracking	
Project Overview	
Cross-functional Project management	
Assignment tags	
Pros	
Cons	
Additional Notes	

FIGURE 8.1
Analyis grid

Due to their organizational system, the team was quickly able to narrow the knowledge management choices from ten to three: Todoist, Trello, and Monday.com.

"Both Trello and Todoist have free versions," Michelle commented. "And even their paid versions are relatively inexpensive. John will like that."

"True, but I think Monday.com offers us the most features to cover the needs of the entire company," Clara commented. "I think we should pick between Trello and Todoist, but keep Monday.com on the list. We can show John the difference between a paid version and a cheaper version. He can choose from there."

"I think we should take that one step further," Michelle said.

> If we truly think Monday.com is the best, then that's what we should write our proposal for. We can also share a write up with the cheaper option so he can see the choices, but I want to place our best foot forward.

The rest of the committee agreed with Michelle and Clara's assessment. Trello operated similarly to a Kanban board, something both John and Michelle were familiar with from their recent visits to Lean conferences and workshops. In Trello, collaborative boards can be created to track the flow of tasks and projects. They can be set up in any way the team sees fit, but the company would benefit most from the To-Do, Doing, Done model of a Kanban board. The committee agreed to present Trello as the second option for John to consider.

"As you know, I recently went to a Lean conference, and I have to admit that I do like the Kanban board nature of Trello," John offered. "The fact that you can create cards for individual tasks and track who worked on each one is a great system. And it's simple. I like simple."

"I agree with your assessment," Michelle commented. "I also like Trello a lot. It's what I use in my personal life, and I can see it benefitting the company in many ways."

"I sense a 'but' coming on," John observed.

> *But*, just because I like Trello a lot personally, doesn't mean I think it's the right fit for the whole company. After taking a closer look at Monday.com, it feels undeniable to me. It offers real-time data collection, elaborate and colorful project timelines, team progress overviews, individualized workspaces, collaborative workspaces, a calendar view, and its own Kanban view. I know you like simple. I also know that the price tag on Trello is a lot more appealing. I think we could get away with the standard version of Monday.com for now. We have thirty employees, which means we would only be paying $300 per month. I know that seems like a lot for this program, and it is, but I do think it's the right direction for the company,

Michelle said tentatively.

TAKING RISKS

John nodded along with Michelle's assessment but knew that he wasn't quite ready to make a decision yet. He wanted to look over all of Michelle's notes, the spreadsheets the committee made, and the cost/benefit analysis Michelle drew up for him to review. According to Michelle, the company was losing an estimated $10,000 per month in productivity caused by errors in communication, failure to track and meet deadlines, and lack of collaborative access to knowledge management across departments. While it would be impossible to address 100% of the productivity issues with a new system, Michelle estimated that the company would be able to increase productivity by an estimated 70–80% with Monday.com despite its high-ticket price.

He turned his attention to Michelle's reports on Trello. While the highest ticket price for this system was only $10 per month, Michelle estimated that it would only save about 30–40% of the company's productivity due to its limited features. In addition, Trello was an older system that was slowly being phased out by newer programs like Todoist and Monday.com. While the difference in savings was only $100–200 a month, Michelle predicted that Monday.com's productivity savings could grow over time, while Trello was likely to be phased out. If that were the case, the entire company would need to learn a new system all over again.

John looked over the current quarter's numbers. It looked like he would have enough money to spend on Monday.com's system if he wanted to use it across the company. It was a bit of a risk, however. What if the company didn't recuperate the money that Michelle and her committee predicted? What if the departments that are already using their own knowledge management systems reject this new system? For John, all of these seemed like big risks. Not to mention, this is John's company. Shouldn't they use the program that John likes the most?

John leaned back into his chair with a huff. This was a lot of brainpower for something he thought would be much simpler.

"Do we even need this right now?" John thought to himself. "I mean, things aren't going that badly. We can probably put this off a while longer."

He knew his company needed to make this move eventually, but did they need to make it *now*? Yes, there were a few bumps in the road, and they were losing out on a little bit of profit, but was it enough to take this kind of risk?

Feeling torn, John decided to call his mentor for a bit of advice.

"I feel like we've been taking a lot of risks lately. I realize this isn't a huge investment, but I don't want to take any more unnecessary risks," John said after explaining his situation.

"Well, how have the risks you've taken in the past few months been going?" Andrea asked.

John paused to think about some of the bigger risks he had taken. First, he started the learning and development budget. Michelle had taken immediate advantage of it and brought back a wealth of knowledge that made immediate positive impacts on her department. She was also able to carry over some of that knowledge into her work with the Knowledge Management Committee. He took a big risk in implementing the AAR process, especially as it pertained to feedback about himself. However, the AAR process had already proven to be a large success in various pockets across the company, who reported feeling more motivated and energized to take on projects.

"I see your point," John chuckled.

> Taking risks isn't only about trusting your employees. It's about trusting yourself too. You're smart to analyze whether or not this risk is right for your company, especially if you are worried about budget or taking on too much at once. But if the project looks viable after all of the analyses are complete, then you have to trust yourself to make the best call for your company,

Andrea said. "**A company with ideas and talent is nothing if it can't take the risks necessary to implement the ideas and leverage the talent of the company**. Does that make sense?"

"It does," John replied. "Thank you, Andrea. I have a meeting to set up."

The next day, John called Michelle into his office. Together, they reviewed all of her documents once more. John asked a handful of follow-up questions about an implementation plan and requested a soft rollout.

> Due to the cost to the company, this one is a bit of a larger risk for us. I think you've done an incredible job with the committee and I trust your instinct. That being said, I saw that the price is different based on the number of people using the system. What I'd like to do first is a soft rollout. Let's set up

> the system in one department – preferably yours – and monitor its success. I think this system has a ton of potential, and I'd like to see that demonstrated before we roll it out to the entire company,

John explained.

Michelle was elated. She understood John's desire to take the process slow while still trusting her committee's instinct on the best system to implement. After the testing period, the Knowledge Management Committee's instincts proved true, and Monday.com was implemented throughout the company.

QUESTIONS TO CONSIDER

1. How do you prioritize your company's goals?
2. How do you recognize the talent on your team?
3. How do you leverage the talent on your team?
4. What systems do you have in place for knowledge management?
5. How do you ensure that new roles are clear, well defined, and achievable?
6. How do you assess your company's needs?
7. What drives change at your company?
8. How does your company communicate expectations?
9. How do you assess risk-taking?

9

Servant Leadership and Innovation

Alex took in a deep breath as he passed the rose bushes that lined the side streets to his favorite pub. The bitter cold of winter was finally beginning to break, and Alex was eager to pull out his spring jacket to meet up with Michelle and Mary Lynn. In the past few months, John continued to send Michelle to Lean workshops and conferences. He was grooming Michelle to be the Lean expert for his company to enhance morale and bolster productivity. Michelle seemed genuinely excited about the progress she was making at TrinoTech, and she invited Alex and Mary Lynn to the pub to celebrate.

Alex bounded through the doors and made a beeline to the trio's regular table. After a round of drinks, Michelle finally shared her news.

"If you had asked me what I wanted to do a few months ago, I would have told you that I was going to start looking for a new job," Michelle said.

> As you both know, working for John was very difficult, and I didn't think I would be able to overcome his working style. However, these past few months have been a complete turnaround for him – and the company. Since I implemented the new Knowledge Management System, our productivity has increased by almost 130%, and the research and development team has been on fire coming up with new ideas.

"The anticipation is killing me, Michelle," Mary Lynn chimed in. "What is the big news?"

> Well, John said that he's impressed by the work I've been doing with the company. He told me that he wanted me to become TrinoTech's Lean expert, which is why the two of you have been seeing so much of me at these conferences. Yesterday, John pulled me into his office to tell me that he wants to promote me into a brand new position: Director of Continuous Improvement!

DOI: 10.4324/9781003318354-12

"That's amazing!" Mary Lynn said.

"Congrats!" Alex praised. "That's so exciting! What does that position entail?"

> I think there are still a few details to be ironed out, but essentially, I will be in charge of standardizing operational processes across the company based on Lean strategies. I will also train all current and new employees on Lean initiatives, conduct regular evaluations of the processes already in place, and work with department heads to develop improvement plans for each area. It's not a role I ever thought I would fill until I started taking some of these Lean classes. I've really been enjoying it, and my talents lend themselves to this kind of work,

Michelle explained. "I'm just excited for the opportunity, and I appreciate both of you for sticking by me these past few months."

"It's right up your alley," Mary Lynn commented.

> This kind of role sounds like it was designed for you. I'm happy to hear that John has been making so many improvements. At first, I was scared that they were just temporary, but it sounds like he's working hard to make these changes permanent.

"I'm happy to hear that, too," Alex laughed. "I didn't think it would ever happen. I left before I gave him a chance. I'm proud of you, Michelle. You put a lot of work into this company, and you deserve the benefits you're getting back."

"Oh, I remember when you left," Michelle commented. "I thought I was going to be right behind you. I think a lot of people were contemplating exit strategies back then. I think that you inspired some of this change in a lot of ways."

"Me?" Alex asked. "How so?"

"If it wasn't for you leaving, I don't know that John ever would have reached out to Andrea for help," Michelle said. "Without Andrea's advice, I think John would still be stuck where he was, and I would be long gone by now."

"Andrea is amazing," Mary Lynn commented. "I think she does a good job helping people understand what they need without telling them what to do explicitly."

SERVANT LEADERS

"She's a good servant leader," Michelle added.

"What do you mean?" Alex asked.

> At one of the recent Lean workshops I attended, they mentioned that Lean leaders also need to be servant leaders. Most people in the room were puzzled by this notion because they related the phrase "servant leadership" to the non-profit sector. The instructor was able to help us understand that servant leadership is more than that. It's clear that Andrea **uses her power and presence to help the people she leads**. In this case, that's John,

Michelle explained.

"That is true. She is always **looking for ways to use her influence to help our team grow and develop**," Mary Lynn agreed.

> Which relates to the **core principle of Lean: to create a learning organization**. Andrea seems to do this with everything she touches. One of the first things I noticed about Andrea was her incredible ability to listen actively. She doesn't just listen; she **listens with intention and without judgment**. You can tell that she puts all of her senses into listening to the person she is speaking with. This ability doesn't come easily to everyone. I know that it didn't for John, but I can see that he's working on it. What I love about her listening skills is that she can make you feel so good about speaking with her that it inspires you to be a more intentional listener yourself,

Michelle explained.

"I would agree with that," Mary Lynn said. "I know that my listening skills have gotten way better since I've worked at her company, and she's never led a workshop on listening."

> She doesn't have to because the benefit of being a good listener is that it often means you're also a good communicator. Andrea knows how to empathize with others. This is a critical piece of servant leadership as it helps people through tough problems or decision points,

Michelle continued.

FIGURE 9.1
How can I help?

> She is phenomenal with empathy. Any time we need to gather information to make important changes to a process or system, Andrea can do so without ruffling any feathers. I feel like a big piece of this is because she **always listens to understand first**. She **doesn't interrupt or interject with her own opinions until she completely hears out the other person**,

Mary Lynn explained.

"That is servant leadership," Michelle said.

> **Servant leaders are self-aware and mindful of how their authority impacts the decision-making process of their team**. They also **allow room for themselves to be vulnerable and admit when they are wrong or don't have all the answers**. Demonstrating these characteristics allows other members of the team to voice their concerns without fear of judgment or retribution. This is especially critical in a Lean environment where learning and growth are valued over mistakes.

"That makes a lot of sense," Alex said. "Marcus is kind and caring in almost everything he does, but don't let that fool you. He is also incredibly talented in identifying the needs of our organization and helping push us in a successful direction."

> Yes. That's another **key attribute of servant leaders: they are very good at identifying needs**. This usually stems from their strong desire to improve both the quality of the products they are helping create and the quality of the work environment their employees experience daily. **By creating a work environment that is aware of employee needs and open to improvement suggestions, servant leaders set themselves and their companies up for success,**

Michelle continued.

"It sounds like it all connects to that core principle of continuous improvement," Alex recognized.

> It does! Creating a culture of learning and continuous improvement is both deeply challenging and incredibly rewarding. **Servant leaders truly understand that learning is critical to their company's ability to grow and thrive in their market**. It's not just something they say or do arbitrarily, they embed their love of learning and growth into the core of their daily operations. **Instead of shying away from mistakes, they lean into them as opportunities for their employees to learn and develop new talents,**

Michelle said.

5E'S OF LEAN INNOVATION

Envision

"This reminds me of the Five E's of Lean Innovation," Mary Lynn interjected.

"Oh boy, here we go," Michelle laughed.

"What?" Mary Lynn scoffed. "It's true. Andrea helped John walk his path through each of the Five E's, starting with **Envision**."

"That one *was* a bit rocky at first," Michelle commented.

> You're right. It was, then Andrea stepped in and asked him to think about the culture he wanted to create. She connected him to Marcus, who challenged him to think about his mission and values. From there, he was able to develop goals and move towards a positive direction,

Mary Lynn countered.

"That's true. Most of the actions he's taken since his session with Marcus have been extremely mission and value-focused." Michelle took a swig of her beverage.

"Which is exactly how it should be," Mary Lynn continued.

> **Meaningful change doesn't happen without a clear vision**. John's only vision for a long time was to simply "be innovative" without truly understanding what that meant. He needed to build a vision that communicated why his methods were the correct direction for the company. Before that, the company worked without purpose. Since then, the company has worked towards new goals and outcomes.

"I would say you hit the nail on the head there," Michelle commented.

"That's because he was able to envision a lucrative future for his own company," Mary Lynn finished.

Engage

"You're right. Once he had the vision, he was able to get more of us on board with his thoughts and processes," Michelle said.

> That's the second E! He was able to **engage** you around his vision. Before he spoke with Andrea, it was the John show. He made all of the decisions without any company input and expected everyone else to fall in line,

Mary Lynn began. "Even after he created his vision, he tried to do everything himself, but Andrea helped him see the benefits of having empathy for his employees."

> It did take him a little bit of time to come around to the idea, though. I remember that terrible meeting vividly. No one understood what he was trying to tell us. The only thing that seemed certain was that all of us were going to be taking on more work than we had signed up for – and we lost two good people over it.

> Then, Andrea stepped in to help him see that **employee input was important**. She told him to design team meetings that encouraged employee feedback. For Andrea, this is the best way for her to build a collective understanding of both employee and customer needs. She takes it one step further, **and participates in Gemba walks around the company to observe her employees and then design goals that align with customer and employee needs**,

Mary Lynn said.

> Which is how John is now. He loves bringing people in to hear their opinions on the way things are going with the company. **He encourages thoughtful reflection and questions so that all of us can learn from one another**,

Michelle said.

"See, that's the piece that was missing for me when I was there," Alex jumped in.

> I never felt like TrinoTech had any psychological safety. I worried that my ideas would not only be turned down but would be highly criticized as well. Honestly, it felt safer to keep my mouth shut and follow orders than it did to speak my mind.

"I'm sorry you dealt with that," Michelle said, "but it is true. TrinoTech went from having zero psychological safety to having a ton. Now we all work together to discuss new ideas and try new things."

Explore

"That sounds a lot like the third E: Explore." Mary Lynn giggled. "Do you see how all of these things fit together and then flow from one another? Without him engaging employees, the discussion piece wouldn't have been possible."

"Further, the more he included us in these discussions, the more he trusted us to unpack the problems we were discussing," Michelle said.

> Now, before we reach a consensus, he asks each of us to do independent research into the problem and come to the next meeting with fresh solution proposals. **It's been both effective and rewarding to sit down as a group to**

thoroughly define the problem and then have a bit of a break to think it through and investigate on our own before we jump into problem-solving.

When we think about that in a Lean context, it becomes even more evident how important the exploration process is. Virtually everything we do in a Lean environment is exploration-based: Gemba walks, kaizen, total productive maintenance, all of these things require a curiosity that enables productive action and continuous improvement. Instead of looking at something and making a knee-jerk decision, **the Lean process slows us down, encourages us to ask meaningful questions, and helps us explore the right problems. There's nothing worse than chasing down the solution to a poorly-defined problem**,

Mary Lynn said.

That sounds like John's new motto. Any time someone brings up an issue with a project or a snag in one of our systems, John's first reaction is to ask questions and engage them with structure problem solving using our new A3-based methodology. A year ago, his initial reaction would have been to offer up a solution for solving it. Now, he wants to make sure the problem we are trying to solve is defined well first,

Michelle added.

"I have to say, I really can't believe that this is the same person who was leading that company when I quit a little over a year ago," Alex commented. "It sounds like he's trying hard to bring people into the fold and empower them to act."

Enable

"Empower is similar to enabling, which is the fourth E!" Mary Lynn said.

I am floored at how perfect of an example John is for this concept. Not only does this flow nicely from exploration to enabling, but it sounds like once he mastered the envisioning and engagement pieces, these latter strategies flowed naturally. Michelle, I remember you saying a few weeks ago that he started putting money from the budget aside to help with a few of your projects. Is that true?

It is. He's been great with offering to fund programs. That's not to say that he just hands money out to anyone, but if he feels that the program being discussed

> has the potential to add value, he's been less shy about backing it financially. The other thing that's been nice about this process is that he stays engaged throughout the meeting without taking over. He provides commentary and feedback, but never tries to steal the limelight from anyone else in the room,

Michelle said.

> **Enabling people to act requires guidance and leadership, but it would wither if he tried to hijack ideas or pull the reigns away from someone else.** I'm really glad to hear he's not doing that anymore. I know that was the one big piece to his leadership puzzle that was missing for a long time,

Mary Lynn finished.

"I think he struggled with his notion of what it meant to be a good leader. Once he realized that we respected him just as much when he took more of a back seat, he was much more productively involved," Michelle said.

> For example, when he asked me to lead the Knowledge Management Committee, he encouraged me to find the right people for the committee. He didn't pry or micromanage my decisions. Though he did give me a few good suggestions, it was nice to know that he trusted me to find the right people. Clara, one of my choices, ended up offering incredible insights to the project and helped us find the best possible product for our company to work with. John asked her some challenging questions at times and offered his opinions, but he never stepped on our toes, and I think we ended up with the best product because of it.

"It seemed like John would never come around to that kind of thinking while I was there," Alex said. "To see that he's been able to do so much in just under a year is a true testament to him."

"It's a testament to Andrea, too," Mary Lynn argued. "Without her guidance, I'm not sure John would have been able to see the light."

The trio laughed together as they reminisced a bit more on John's company's progress since employing Andrea to help him overcome his hurdles. They passed stories of both gratitude and awe back and forth between one another, all in recognition of Andrea's hard work as a servant leader in an innovative field.

"I wonder how long it would have taken to get here without Andrea's guidance," Michelle wondered.

Evaluate

"You mean, you wish there were a way to **evaluate** your progress against what likely would have been very little progress?" Mary Lynn emphasized "evaluate."

"I am sensing that evaluate is the fifth E," Alex asked.

"That would be correct," Mary Lynn continued. "This is the last one, I promise. Then we can talk about something else."

Alex and Michelle made eye contact and rolled their eyes together. Mary Lynn decided to ignore them.

> Evaluate is one of the most important E's of Lean Innovation. As you both know, the Lean industry puts a lot of emphasis on evaluating processes, strategies, products, etc. **The only way to know whether to continue with, edit, or completely walk away from a project is to evaluate the project's outcomes**. Without evaluation, it's impossible to know which direction is best for a company to move in. Take TrinoTech, for example. John hasn't been able to sustain his level of effectiveness by constantly doing the same thing. He had to adapt.

"That's true," Michelle said.

> He's been very good at evaluating everything that we do. We have even adapted the AAR process to encompass a broader range of topics so we can continue to grow in areas like customer relations and sales competition. The process adjustments and progress we've made this past year are based on data we've collected either internally or from our clients.

"Now, that's innovation," Mary Lynn said. "It sounds like John is focusing only on value-add activities. Would you agree?"

"Yes. I would say that's accurate," Michelle answered.

> Our evaluation processes lead to changes that enhanced our product or service, eliminated waste, and were customer-approved. It's been incredible to see our growth. The past two quarters have been especially impressive. I think there's a lot to be said for the new knowledge management system helping us keep track of communications we send to the sales team. They have been kicking butt over there.

"That's the **purpose of a good evaluation. It's supposed to lead to stronger sales numbers, company alignment, and overall boosts in morale and productivity,**" Mary Lynn said. "So, there you have it. It sounds like your company is really taking these Lean lessons seriously – and it's paying off."

QUESTIONS TO CONSIDER

1. How do you demonstrate servant leadership?
2. How do servant leadership and Lean leadership overlap at your company?
3. How do you enable continuous learning?
4. In which of the 5E's is your company strongest?
5. Which of the 5E's could your company improve upon?
6. How does your company communicate its vision?
7. What does your company envision for its future?
8. What do you envision for your future?
9. How do you engage the talent around you?
10. How does your company engage its talent?
11. How does your company engage folks in decision-making processes?
12. What helps your company properly define problems?
13. How does your company empower its workforce?
14. How can you improve upon your evaluation process?

10

Continuous Improvement Mindset

10 Years later

Damon Gilbert had been working at TrinoTech for a little over three years and thoroughly enjoyed his work. He knocked on a door labeled Vice President of Learning and Development before stepping inside.

"Good morning, Michelle," Damon said. "Here are the reports from this week's Gemba walks."

Damon handed the documents to Michelle for her review. This was fairly routine for the two professionals as Damon was the senior director for Learning and Development for the production side. He worked closely with Michelle to identify areas of growth, professional development opportunities for his teams, and processes that could help the company reduce waste. Their partnership on Lean initiatives over the past three years allowed John to pursue other company goals. They were recently able to make enough capital to open another branch of their business.

With a new start, John worked closely with Michelle to ensure every aspect of the new branch started with Lean in mind first.

"The most important thing for this company is to continue down the path we started to carve out ten years ago," John explained.

> Looking back, I still can't believe it took me so long to come around to some of the more progressive ideas that brought us to where we are today, but we can't afford to lose that momentum with the opening of the new branch. I think we should start with revamping our virtual training videos to be specific to the needs of the new branch.

DOI: 10.4324/9781003318354-13

INTERNAL LEARNING

Rethinking Development

For a long time, John and Michelle worked hard to offer a variety of learning experiences. Michelle's team was always on the lookout for new conferences or workshops that were providing educational experiences on topics relevant to TrinoTech. Michelle created a spreadsheet, organized by topic, location, length of program, and price. She and her team hand selected programs they felt would benefit both the employee and the company most.

For the first few months, folks were delighted with this procedure. They had consistent and constant access to professional development opportunities in their area of expertise. They could see when the program was taking place, how long it would take them away from work, and the cost to request from TrinoTech. The program offerings were robust, and each employee was actively encouraged to attend at least one professional development opportunity per year. Most folks at the company were both excited and energized by the influx of new opportunities. Company morale soared as employees rushed to take advantage of the professional development funds being offered.

It didn't take long, however, for the morale to slowly taper back down. The number of employees taking time to pursue professional development opportunities flat-lined, and the programs most sought after also became the most expensive. It became abundantly clear to both John and Michelle that the way employees were pursuing development at TrinoTech was not in line with their original vision.

"Is this what development should look like at our company?" Michelle asked John after reviewing the lowered morale numbers.

"That's a great question," John responded. "This is the question we should have considered when we started the Learning and Development fund. What does development mean at TrinoTech?"

"I will work with the committee to draw up a proposal for what development should look like and how we should pursue it," Michelle said.

Michelle had created a small Learning and Development committee in the months since taking on the new role to help her curate new opportunities for employees and vet the relevance of previously unidentified opportunities. For weeks, Michelle's team worked tirelessly to answer the question of development at TrinoTech. Unanimously, they decided that external opportunities for professional development should not be the sole way for employees to grow. If they truly wanted a continuous improvement mindset at the company, there had to be more than external conferences. Continued learning needed to be deeply embedded in the culture of the company daily.

Michelle and her committee decided to define development as "an ongoing, employee-centered approach to continuous learning where dedicated resources and strong expectations are laid out to provide regular learning opportunities." With this new approach, employees would be encouraged to seek new ways of learning, without being restricted to one large conference or two small workshops per year. With this in mind, Michelle's committee identified five key areas for internal learning to take place: licensing and learning partnership, online learning, employee knowledge sharing, mentorship, and think tanks.

Continuous Learning in Onboarding

Before Michelle and her team could address any of the five areas for internal learning, she knew they needed to address their archaic onboarding process. As it stood, new professionals were onboarded almost entirely by HR. Michelle had been frustrated with the process since her first day on the job, remembering how impersonal and cold it had felt to join the team. With the tides turning at TrinoTech, Michelle wanted to take advantage of instilling their new definition of development into the learning process.

Michelle developed a five-step plan to revamp the old ways of onboarding into an invigorating and motivating experience. Her goal was to identify each step of the process and make the process both Lean and personal.

First, Michelle wanted the onboarding process to start before the new hire arrived for their first day. Instead of coming to the building on the first day to fill out paperwork, she proposed moving paperwork to an online portal that new hires could do on their own time *before* their first day. Michelle reasoned that this proposal was in-line with the company's efforts to eliminate waste. HR professionals wouldn't need to waste precious time waiting for a new employee to fill out paperwork in their office, nor would they need to print any paperwork out in the first place. This would free up time for HR professionals to focus on the individual and not completing a series of HR tasks.

Michelle's second step was to make the process more personal. Michelle vividly remembered her first day: It had been raining hard that morning, and she walked in sopping wet. No one greeted her, nor did they offer any kind words to make her feel more comfortable. She waited almost 30 minutes for the HR professional who seemed so overwhelmed that she barely noticed Michelle's wet state. When she did, she merely acknowledged how uncomfortable it must have been and then continued with her 90-minute HR presentation. It was horrible.

FIGURE 10.1
Bad experience

Instead, Michelle wanted the HR process to be more personal. Before a new hire walked through the doors for their first day, she proposed sending a personalized welcome letter. This letter would arrive a few days before their start letting them know that the rest of the company was eager to meet and work with them. On their first day, HR would meet them at the door and walk them to their cubicle, where a small company gift would be waiting on their desk. Again, this gesture was meant to send the message that TrinoTech goes above and beyond for their employees. Then, HR would walk the new person around the company and introduce them to the team's key members. At this point, the new hire would learn who their resources are and how to find and contact them. Michelle proposed that

HR should be thorough during the introductions and encourage the new hire to write down names and positions for future collaborations and learning initiatives. Finally, Michelle proposed scheduling a team lunch or coffee outing for the new hire and their team to get to know one another outside of the work environment. The goal of this outing would be for the team to answer any questions about the company and clearly articulate company culture.

FIGURE 10.2
Good experience

Michelle's third step was to include a special presentation that focused on the company's vision and mission. It took months for Michelle to grasp the company's purpose and goals. Even then, John would change them every so often, which caused more confusion and drove down company engagement. Instead, Michelle felt strongly that the employee's immediate supervisor should sit down with the new hire to review the history of the company. She also wanted the supervisor to explain not only that the company was growing, but *why* the company was growing. From a continuous learning mindset, Michelle felt it was important for new hires to know why the company prioritized certain projects and opportunities to maximize their success and growth. This third step provided a rich opportunity for the new hire to learn more about the company's purpose. It also provided another personal touchpoint between the new hire and their supervisor.

Of course, no onboarding process would be complete without relevant training. Michelle wanted to work with her Learning and Development team to come up with a vigorous training for new hires that outlined information on company dynamics, job performance expectations, processes and procedures, and Lean initiatives. Her plan was for each topic to be covered by another member of the company to provide additional touchpoints for new employees. For Michelle, one of the biggest things that defined her success at TrinoTech was the connections she made with fellow employees. If interested parties delivered each of these training topics, she felt strongly that the new hire would feel more connected and excited to contribute to their team.

The last piece of Michelle's onboarding plan was simple. She proposed a one-month follow-up with all new hires. This follow-up was not meant to be part of a performance review, but rather to check in and assess how the employee is adjusting to the new work environment. This would allow TrinoTech to assess any gaps in their training process and help the new hire feel heard and cared for.

Michelle's five-step plan was designed to build morale, embed continuous learning and development into the onboarding process, and help new hires develop personal connections to their teams and their work. By approaching this new process from a Lean perspective, she was able to build something that was both sustainable and rewarding for everyone involved.

Licensing and Learning Partnerships

With a new onboarding process ready to roll out, Michelle decided to turn her attention to the five internal learning initiatives she and her committee had identified. Shortly after Michelle became director of continuous improvement, TrinoTech started to run out of money for professional development opportunities. Many employees were using those funds to seek external opportunities for professional development at conferences and seminars across the country. The expenses for these opportunities began to increase, and it was difficult for TrinoTech to keep up with company demand from a financial perspective.

Michelle sat down with her Learning and Development committee yet again to brainstorm potential cost-effective opportunities for individualized learning and growth. They spent the first meeting trying to identify the root of the problem. They looked back through all of the recent professional development requests to identify any potential patterns in the topics people were seeking out for growth. Overwhelmingly, folks were attending workshops and seminars that dealt with Lean learning and innovation. While there were a few outliers on product development or sales tactics, people were starting to invest in John's efforts to make the company more Lean and innovative.

"This is great news," Damon said.

> Maybe we should consider licensing online Lean learning initiatives that already exist. Instead of sending people to new programs and workshops all over the country, it might benefit us to partner with a Lean training company and encourage our folks to attend programs through this partner.

"I like this idea," Michelle said.

> Since it seems that Lean learning and innovation are the topics people are most interested in, I'm also wondering if it makes sense for a Lean consultant to set up workshops on-site once a month for anyone to attend. This could be added to that partnership if the Lean programs are local. With John pushing for more Lean initiatives, this might be the best time for us to propose a more robust approach to training Lean concepts across the company.

The Leaning and Development committee agreed wholeheartedly. Michelle knew that it was vitally important for the company to have a strong knowledge base of Lean topics for the Lean processes and procedures to be most effective. Over the next few weeks, Michelle worked with local Lean consultants and Lean learning programs to develop a robust Lean curriculum.

She first wanted to target current employees. She made sure the new Lean curriculum included information on managing through change and focused on the benefits of Lean processes. Her goal was to make sure that all current employees felt confident in their ability to address problems with a Lean mindset.

Her second initiative involved new employees and the onboarding process. While she wanted to allow these folks to dive into Lean learning programs immediately, she was also painfully aware that there might be a learning curve for folks who had not come from Lean environments. She decided to revisit the onboarding process she had developed and decided that it would be best for all new hires to go through a Lean basics course offered through the Lean training company they were hoping to hire.

With the help of her Learning and Development committee, they pulled their resources to better understand incoming employees' needs, especially as it pertained to the new branch. She wanted to make sure that Lean culture was apparent to everyone as they began their new duties, making sure that the learning was relevant to both senior and junior employees. At the end of their research inquiry, they felt that the Lean Learning Center was the best option for this new initiative. The Lean Learning Center had the ability to offer local, in-person workshops and classes, a series of hybrid programs, as well as easy-to-digest online content that folks could take at their own pace.

Michelle and her committee took the proposal to John, adding that they would need half of the Learning and Development funds to pay for this new initiative. While they recognized that it was a healthy sum of money, they also articulated that the company would potentially save money. Most employees were already interested in pursuing programs based on Lean and innovation initiatives. Now that the partnership between TrinoTech and the Lean Learning Center existed, most employees would prioritize these programs over others.

Knowledge Sharing

The learning partnership wasn't the only aspect of the programs that Michelle and her team developed. After seeing the successful results of the learning partnership, Michelle realized that creating a robust internal learning program would also greatly benefit company dynamics.

Once again, she brought her team together to propose employee-led discussion-based seminars. She wanted to offer an education series monthly to target two distinct populations at TrinoTech: company experts and folks who participated in various learning initiatives.

During the onboarding process, Michelle noticed a number of company leaders step up in a big way to provide knowledge to new hires. A few folks put on Q & A sessions, some made mini PowerPoint sessions, and others provided hands-on training opportunities. It was inspiring. It reminded Michelle that the leaders at TrinoTech were well versed in their areas of expertise, and some of them thrived in a teaching environment. Michelle approached a few leaders with the idea of hosting quarterly lunch and learns. During this lunch, they would teach a short session and then take questions about their specific area of expertise. The lunch and learn would be open to anyone in the company. Michelle was shocked by the response. Ten of her departmental leads agreed to host a lunch and learn.

The second population with knowledge to share were the folks who took advantage of learning opportunities through the company. The idea was to create a process by which any person who used Learning and Development money for an external professional development opportunity would be required to bring the knowledge back to the company intentionally.

Her committee loved the idea. Since TrinoTech was paying for employee development, Michelle believed the information should benefit the company. They decided to make it a requirement of all Learning and Development opportunities. The expectation was that each person would prepare a workshop, seminar, or detailed resource document for the rest of the company to learn and grow from. These folks were also able to develop "lunch and learns" or "coffee chats" to disseminate their knowledge to others.

Both initiatives were a huge success, pulling in a diverse subset of the company to grow and learn together. Employees were able to gather knowledge, ask questions, and share ideas with the intentional goal of making the company a more collaborative and innovative work environment.

Online Resources

The knowledge-sharing initiatives were so popular that Michelle's committee began to record them. After the programs were over, Michelle would upload the recorded conversations to a company-wide Box file. Anyone who couldn't make it to the program could access the conversation and learn from the information provided. Over time, Michelle's committee noticed that the online portal had become increasingly popular.

> I think we need to consider making this platform more than just a place to store lunch and learn videos. Our company's culture is one of continuous learning and innovation. People are hungry for more knowledge about what we do and want to learn better ways to do it,

Damon, one of Michelle's committee members, said.

"I agree," Michelle said.

> I'd love to see more resources in our Box folder. We should sit down and map out what we want our online learning resources to look like. We have a strong learning partnership with the Lean Learning Center. Still, we need to provide information around trends in our industry, cutting edge research in the materials we are using, or general guidance for things like self-care and resilience. I think Damon makes an excellent point; we can offer a lot more than just our lunch and learn recordings.

FIGURE 10.3
Choose wisely

Over the next few months, Michelle's team worked hard to identify the kinds of resources the company wished were available to them. They held meetings with various stakeholders to identify key topics in the field, sent out surveys to identify the needs and wants of the employees, and brainstormed ways to incorporate company-specific initiatives into the resource compilation. The project was a beast. Together, they created the first-ever interactive Lean Learning and Innovation Online Resource Center. Through the Box platform, they developed folders labeled for various resource categories such as onboarding, innovations in the valve industry, resilience and wellbeing initiatives, innovative techniques in technology, and creativity. Each folder was broken down into subfolders, some containing videos and Ted Talks, others containing empirical research or popular media articles. Michelle's team rolled out the online resource center with one request: add anything you read or watch that would benefit the growth and development of the company.

The resource center was exactly what John was hoping for as the new branch opened. He asked Michelle to add resources for implementing Lean learning and innovation during the development of the new branch. Luckily, Michelle was well versed in addressing such needs. She pulled together her team and assessed the needs that the incoming team would have. They would need both an onboarding process for Lean and innovation and resources that outlined what it meant to be a part of the new branch. John wanted these resources to inform the new team members that their very existence demonstrated the company's innovative nature.

Understanding John's vision for the new branch clearly, Michelle put her team to work. They worked together tirelessly to compile resources for new hires while also providing progressive guidance to the onboarding process. This project inspired Michelle's team to develop their own set of resources outlining a few key aspects of being a TrinoTech employee. They added a self-playing presentation on the mission, vision, and goals of the company. They produced and published fact sheets for each department describing both how and why the department was started. Finally, they included a recording from John that outlined the company's history in a fun and engaging way.

Since creating the first online resource center almost ten years ago, TrinoTech has added hundreds of resources for employees to take advantage of in their own time. Not only did this reduce the budget necessary for professional development, but it drove morale and productivity. Everyone was treated both as learners and as contributors.

Michelle's team has continued to notice big improvements since the first iteration of the resource center. John mentioned that the Learning and Development team asked for a slightly smaller budget for the third quarter and wanted to know why. After doing a little digging, Michelle's team realized that folks would rather invest in the learning offered through the company than seek additional learning elsewhere. With the addition of the Lean Learning Center partnership and the Online Resource Center, Michelle noted a substantial increase in the sense of loyalty and trust that most folks feel across the company.

Mentorship

Michelle's motivation stemmed from wanting to create a strong mentorship program. They had tried a few times to create one from scratch, but something about the program never stuck. Michelle's intention with the knowledge share program was to start an evaluation process. Her goal was to see what experts within the company were willing to teach others about their topics, and who was hungry to learn more.

Michelle tracked the number of repeat attendees at similar workshops across the company. She wanted to separate the people in the company who had a deep love of learning about new things in general from those who were interested in learning about one topic on a deeper level. From there, the ladder group was added to a list of folks in need of a mentor within the company. That list was cross-referenced with a list of folks who were willing to offer education and mentorship.

This was where Michelle shined the most in her Learning and Development role. She was incredibly perceptive to the needs of her company and was able to identify the best ways to bring people together in the name of continuous learning and growth. Michelle understood that the best way to create an environment of continuous learning was to consistently offer opportunities for people to learn and grow – and that it would require both funded and unfunded opportunities.

Think Tanks

With this thinking in mind, it didn't take long for John to create his version of Andrea's Think Tanks. Instead of monthly meetings, however, John decided to dedicate an entire department to innovative thinking, with a company-wide invite to any meetings that discussed projects of interest. With Michelle's help, John laid out five ways for his company's think tank to maximize their impact.

First, he encouraged them to **look for similar ideas across the company**. If production was looking to streamline one of their processes, the think tank committee was charged with determining whether or not similar streamlining was a need in other departments. If sales wanted to come up with a more precise language to use when talking with customers, marketing was invited to the conversation to provide input that would benefit both departments.

"When we combine ideas across the company, we are opening doors we never knew existed," John said at the committee's first meeting.

It was true. Once the company was notified that John was looking to combine ideas, opportunities TrinoTech never imagined started popping up, including an opportunity to partner with NASA on an experimental valve for their rockets. This opportunity led to TrinoTech's new department focusing on designing valves for the aerospace industry.

His second recommendation was to **encourage his think tank to imagine an ideal product. What would it look like? How much would it cost? They were required to start with the ultimate goal in mind**. Once they could imagine and describe it, John challenged them to **work the product backward and determine the best course of action for seeing the project to fruition**. Though this was not a standard meeting process, it was their most creative option. Whenever the team felt stuck or tired, they used this method to get their creativity flowing. A few times, it worked, and they were able to create products that once only lived inside their imaginations.

His third strategy was utilized most during the development of smaller projects. It involved **taking an idea from the drawing board to a full-dimensional model quickly. The idea was to catch design flaws, process errors, budgetary issues, etc., as soon in the process as possible. This eliminated time and material waste that would otherwise be used on projects doomed to fail**. This was one of his greatest improvements. No longer were teams chasing down impossible projects and forcing them through the pipeline. Instead, they were able to quickly determine the legitimacy of a project and make the decision of whether or not to move forward faster than they ever had before: no more wasted time, materials, or energy.

The research and development fund began to grow as the think tank continued to produce top-level products. As this budget grew, John employed his fourth strategy and **encouraged his company to network with one another to chase new opportunities**. He recruited Michelle to create an online learning workshop that addressed strategies for networking with purpose and within the company's walls. John's goal was to find ideas vastly different from those the company was already funding and use the research and development funds for these projects. Michelle loved this idea. She had been working on networking at TrinoTech since she first started and had done a variety of independent research to develop the best strategies for finding and maintaining the right connections. The workshop was a success, and new pockets of teams developed to work on high-level collaborations. One of these collaborations led to large improvements in their company's most popular valve, driving sales for that product by 25%.

It was John's last strategy, however, that shocked Michelle so much. Over the years, one of John's key rules was to bring new ways of learning into the company and share learning across the company as much as possible. However, he preferred these interactions to happen in person. He wanted people to take the time to meet, discuss, and experiment before putting anything into action. In-person meetings, however, came at a cost. There were times that key players failed to show to the in-person meetings due to workload issues, and a few times projects were dropped due to the inability to find a good overlapping meeting time.

Seeing this problem, Michelle recommended an idea management software, similar to the knowledge management software that had been so successful at the company years earlier. She did quite a bit of research and found a company named Miro that allowed real-time collaboration in an online platform. It effectively eliminated the need for folks to meet in person so frequently. Instead, teams could form through the software, discuss ideas, create templates, and mind maps, all while keeping up with their own workloads. Of course, this didn't replace in-person meetings as the project began to take concrete shape. However, it did offer folks alternative means for collaboration during the initial phases of a project, which offered the flexibility the teams needed to get the projects off the ground.

FIGURE 10.4
Team work

Struggling to Thriving

John returns to his office at the end of a long day. He logs into his email to review Michelle's weekly reports for the company's learning and development initiatives. Among other things, they measure employee satisfaction and productivity. As he stares at the numbers, he reflects on Alexander's departure from the company and how it catalyzed this uncertain journey.

For months, John struggled with himself and his vision. He was forced to grapple with the needs of the company versus his need to be top dog. He confided in Andrea, a mentor and friend, who offered ideas that were uncomfortable and unfathomable in John's eyes. She connected him to Marcus, another mentor, who encouraged John to follow Andrea's advice with a little help from the magical world of Lean. With Andrea's skills and talents in innovative initiatives and Marcus's expertise in Lean, John had surrounded himself with the best people to help him solve these problems.

John chuckled to himself as he thought of Michelle. She must have thought he was crazy when she first started with the company. He shuddered to think that she may have even considered leaving if it weren't for Andrea's interventions. Now, she was one of the most important people. He couldn't imagine having the same success he was experiencing now without her incredibly innovative mind and drive for continuous improvement. Her talents for networking and love of learning provided the essential building blocks he needed to create a company he only dreamed of having one day. He is eternally grateful for Michelle's loyalty and passion for her work.

He glances down at the numbers again and allows his emotions to bubble. Productivity has gone up another 10% since last week, almost 300% over the past two years. Employee satisfaction, something that John wouldn't have bothered to measure ten years ago, is at an all-time high. His people come into work every day, loving their roles in the company. John has made it his mission to make sure every person who works for him understands their value and recognizes people for their contributions to the company every chance he gets.

John's story wouldn't have been possible without his decision to change, his commitment to making the company better, and his ability to embed a culture of continuous learning into the core principles on which his company operates. Ten years ago, John's company was on the verge of falling apart before his eyes. Today, it's thriving.

QUESTIONS TO CONSIDER

1. How could you improve your onboarding process?
2. What would make your onboarding process more personal?
3. What would make your onboarding process more efficient?
4. What kinds of internal learning take place at your company?
5. How can your company grow its online learning potential?
6. Who could you partner with to further develop learning at your company?
7. What resources does your company currently have access to?

8. How can your company better leverage the knowledge of the folks within the company?
9. How could think tanks benefit your company?
10. What kind of mentorship does your company offer?
11. What shifts need to happen in your company to make learning a higher priority?
12. What does development mean to your company?
13. What does your company do to manage the ideas of its teams?
14. Does your onboarding process include an introduction to Lean thinking?
15. How do folks at your company know that their ideas are valued?

11

The Key Concepts of Fostering Innovation

"It's not that I'm so smart, it's just that I stay with problems longer."
– Albert Einstein

Innovation is not an easy nut to crack. Ironically, Merriam-Webster defines the term simply as "a new idea, method, or device." That's it. So, it begs the question: if innovation is so simple, why do so many people and companies struggle with it?

This is where the complexity and nuance of innovation come into effect. If you walk into any company today, they will likely tell you they strive to be innovative and creativity driven. They will tell you that they make efforts to identify the best new ideas and put processes in place to bring them to fruition.

The issue isn't the *desire* to be innovative. The issue boils down to two key things. The first is a company's inability to dig deep and identify the characteristics, resources, and environment that is needed to make a person or company innovative. The second is that same company's mistake in making innovation out to be something larger than it is.

ISSUE 1: DIG DEEP

Some companies are relatively unwilling to do the work necessary to cultivate the people and environment needed to generate innovative ideas. This work includes identifying and recognizing the thought leaders of the company. It means understanding that these people exist at all levels in a company, not just at the top. Frontline employees are sometimes in the best position to offer up new ideas to help drive a company forward due to

DOI: 10.4324/9781003318354-14

their proximity to and knowledge of the work and problems. Regardless of where innovation comes from, it should be recognized and rewarded until it is a standard operating procedure. It should be something all members of the company strive for daily, not just when there's extra time or when there's a problem.

It's important to note and identify the characteristics of innovative minds. Throughout this book, you were introduced to an overabundance of innovative characteristics. From good note-taking and networking to experimentation and mindset, an innovator requires a slew of characteristics to be successful. Most innovators indeed possess a level of creativity and ingenuity that allows them to think and act outside the norm. They are not necessarily the smartest people in the room, nor are they the ones who sit around and wait for an "aha" moment. However, they are the people who show true grit and determination to solve complex problems or develop better ways to achieve goals. They don't give up easily, and they are energized by the intrinsic rewards they receive once they have successfully solved a problem.

Innovators exist at every company, but they need to have access to a variety of resources to develop their skills and feel safe enough to explore. Money is not the only resource needed. While money is important, other resources are equally relevant to an innovator when experimenting with a new idea or concept. Time, access to other colleagues, educational opportunities, growth opportunities, and strong communication are important resources to enable an innovative spirit in every employee. If employees are so overworked that they cannot focus on a new idea, then those new ideas die in their heads before they ever have a chance. Employees need brain fuel in order to survive and thrive. Teams who have access to additional education or growth opportunities often flourish simply from their ability to bring new ideas to their teams. Communication is also key. Whether this is communicating product defects from customer service to production or communicating critical design changes from engineering to sales, communication can sometimes mean the difference between the success and failure of a product. Having clear channels for communicating key pieces of information is a crucial resource for every member of a company.

Innovators shine in environments that enable their skillset. Even the best problem solvers require an environment that cultivates their innovative spirit or risk burning out. Culture change starts with having access to the resources they need to think up new ideas, but it doesn't stop there. Leaders who provide open channels of communication and safe space for their people to take risks provide ripe conditions for innovation.

Not only are innovators persistent, but they are proactive, taking it upon themselves to work through challenges long before they are asked to address them. Autonomy, then, becomes a critical piece of the company's environment. How does the company define autonomy, and what parameters do employees need to follow to start work on a new project or solution? How can other people be involved? Standard processes and procedures should be in place for folks to follow while giving them the space they need to perform at their best.

Leaders should be able to communicate these standards across the company clearly and concisely. When it comes to innovators, however, leaders should be prepared for potential push back. At times, innovators can rub leaders the wrong way, as they tend to challenge the status quo to drive forward progress. Strong and confident leadership is pivotal to run an innovative business. Instead of pushing back against the innovators and their challenges, it should be standard to lean into them. Allow them to explore and experiment with new ideas to make the company stronger. Empowering employees to help in this effort will add to and continuously build an innovative culture.

Beyond clear communication and empowered employees, however, leaders need to be willing to step into the ring with those they lead. This means abiding by the same standards, procedures, and practices as the rest of their company. An empowered employee's energy will only last as long as they see leaders modeling the kind of innovation they are looking for and playing by the new rules. If employees see a leader step out of line, even for a moment, morale and motivation will decline rapidly. If leaders want innovation to become a core value of their company, they need to dive in headfirst and stand on the front lines with their teams.

ISSUE 2: SEEING INNOVATION IN SMALL STEPS

Innovation is simply not as large and grandiose as many people would like it to seem. Though leaders across various industries define innovation differently, they all agree on one common element: innovation is the ability to stay relevant. This doesn't mean that people are constantly coming up with grand ideas that draw millions and millions of dollars into the company. It simply means that the ideas produced and executed by the company maintain or increase a product's desirability in the market.

If you think back to the example of Blackberry and iPhone, you can see where this becomes important. While iPhone was coming out with touch screens and dual cameras, BlackBerry was sticking to physical keyboards and cameras with poor quality. At first, BlackBerry didn't seem to mind their drop-off, sighting that they were meant more for business-minded folks and not necessarily the everyday person. This changed drastically when iPhone started adding business-related apps and software to their design, making them equally competitive from a business standpoint but ten times more desirable from a design standpoint. BlackBerry didn't need to invent the iPhone to stay in business and keep growing; they just needed to do something that kept their product relevant. When they didn't, they faded into the background.

Innovation isn't one big "aha" moment that puts the company on a map, nor is it staying the course and ignoring unique thought. It requires the effort of every company employee – from the CEO to the production team and everywhere in between. Everyone should feel empowered to offer new ideas, experiment when it is relevant, and provide opportunities to learn from potential mistakes.

Innovation is hard. It's even harder without a willingness to change or stubborn leadership that wants all of the glory for themselves. However, when innovation is done correctly, it can open up more doors than just staying relevant.

A side effect is innovation's ability to improve production. Like Lean, innovation seeks to eliminate waste and reduce costs. The less waste and costs a company incurs, the more space and revenue it has to be creative and experiment with new thoughts. Innovation is all about improving through experimentation. The experimentation process encourages people to focus on solutions, improve old systems, or develop new ideas. All of these things lead to an increase in productivity.

Quality and value are essential to the reputation of any company. Products that have a higher quality or add value to the market are more likely to meet customer's needs. Not only will increases in quality and value please customers, but they will likely lead to an increase in sales and profits. When an innovative idea is effective enough, it can also lead to more demand, which means a customer will be more likely to pay a higher price for the product.

Innovation is about keeping up with competition and, hopefully, coming up with a few ideas to keep you ahead of it every so often. When innovation is working at its finest, it helps put companies in the spotlight, boosting marketing efforts and beating harsh competitors. Creating a new brand, developing a catchy ad campaign, or focusing on a quirky aspect of the company are all innovative ways to enhance marketing campaigns.

Arguably, one of the strongest benefits of innovation is its ability to retain motivated staff members. A company's reputation is as good as the people who work for it, and vice versa. The best people in the job market are attracted to companies with a strong reputation for innovation due to their uplifting and stimulating nature. Today's world is filled with brilliant minds looking for a place to expand their knowledge and make the world a slightly better place. Create an environment that is both fun and inspiring, and top-quality talent will find you.

Today, more than ever, our world needs innovators. People and businesses alike must implement and utilize the tools and strategies mentioned

FIGURE 11.1
Run with it

throughout this book to develop thought leaders worldwide. While this book is not the only answer, it can provide some of the insights necessary to create the environments we so desperately need. Through the lessons of Lean, a robust focus on culture and values, and a desire to change and grow, the next phase of innovation could be the revolution our world has been waiting for.

Glossary

A3 Report: An easy-to-read report that conveys the problem and proposed solution for a project or improvement. A waste-free and structured form of report writing that frames how to think about problem-solving, improvement ideas, or a project.

After Action Review (AAR): Planned event, originated by the United States Army, to review and reflect upon the results of simulation missions. The AAR brings intact work teams together to analyze and reflect upon an event to understand what and why things happened during the process and learn from that experience. The event typically asks the following questions:

- **What should have happened (what did we expect)?**
- **What did happen (actual results)?**
- **What is the difference between what should have happened and what did happen (the delta)?**
- **What will we do differently next time (lessons learned)?**

Autonomy: Giving people ownership over their creative process to instill a sense of ownership over one's work. From autonomy grows creativity and the desire to contribute more to the company.

Brainstorming: Occurring in many forms, these group discussions are meant to produce new ideas and solve potentially complex problems.

- <u>Five Whys</u>: **A structured process to dive deeper into any problem to A) eliminate problems and B) use problems to move to another level of performance. Process of asking "why" repeatedly when a problem occurs until the root cause is uncovered.**
- <u>SWOT Analysis</u>: **A structured analysis tool that can be used to quickly assess the current strengths, weaknesses, opportunities, and threats of a process or organization.**
- <u>Reverse Brainstorming</u>: **A structured way to think of a problem backward by identifying several different ways to make a problem**

worse. The focus is then on addressing the gaps in the progress the team identified.

- **Most Ideas Contest: A structured brainstorming activity that places value on quantity over the quality of ideas generated. The more ideas generated, the stronger the ideas become.**

Change Management: The process that guides one's preparation for change. It includes the resources, support, tools, and techniques used to manage people during a major business change.

Coach: Typically, a hired professional who uses structured, formal approaches to assist someone with a specific goal or vision. This relationship is designed around a short-term learning outcome with a focus on boosting performance indicators and improving skills. These relationships have measurable outcomes.

Continuous Improvement: Improving products, services, processes, and procedures in an ongoing fashion and without a clear "end." Continuous improvement means that there is no ultimate destination, just the process of continual growth and development.

Creativity: Merriam-Webster defines it as "the ability to create." In this book, we define it as "the ability to think up new and unique ideas."

Culture: Merriam-Webster defines culture as a set of shared attitudes, values, goals, and practices that characterizes an institution or organization. Largely defined by leadership, it should be something that is easily identified within your company. It will inform the way the company runs and how employees engage with one another.

Culture Audit: An assessment that helps leaders understand the current state of the culture at their company. Aspects of this audit include identifying the messages employees currently receive and the effectiveness in communicating accurate information. It helps leaders identify how and why the culture of the company exists as is.

Gemba Walks: A "go and see" practice to understand the current condition of an organization's value creation process (value stream) in order to develop the team's problem-solving capabilities.

Idea Programs: A structured process for employees at a company to think of and submit new ideas.

Innovation: Defined many different ways. Merriam-Webster defines it as "a new idea, method, or device." In this book, we define it as "the ability to identify unmet needs and introduce progressive change."

Innovation Adoption Curve: This model classifies the adopters of innovative ideas into five main categories:

- **Innovators: These are the first people to try out a new idea.**
- **Early Adopters: These are the highly respected members of society who push the idea forward due to their heavy influence over others.**
- **Early Majority: Well-connected and well-liked members of society who spread the innovative idea quickly.**
- **Late Majority: A more skeptical group of adopters who only come on board well after everyone else already has.**
- **Laggards: These folks are very traditional, actively resist change, and get stuck in their ways. They often adopt an innovation long after it has been available.**

Intrinsic Motivation: The act of doing something without any external benefit. Folks who operate from intrinsic motivation work for the love of the task or project they are executing. They are not motivated by the rewards that come with the completion of that task.

Knowledge Management: A system used to efficiently organize information and resources at a given company. It includes the process of creating, sharing, and using the knowledge of an organization.

Learning Systems: A system used to maintain idea programs, research and development teams, and other innovative projects. It includes processes, procedures, and projects that all contribute to the constant learning and growth of both the company and its employees.

Mentor: A trusted advisor from a professional relationship. Mentor relationships are often built up over time, rely on mutual trust and respect, last for long periods, and focus on building the mentor and mentee relationship.

Mission: Sometimes crafted in the form of a statement, this word refers to a specific task or goal a group is charged with achieving.

Psychological Safety: A concept coined by Dr. Amy Edmondson to refer to the "shared belief held by members of a team that the team is safe for interpersonal risk-taking." It means that team members can speak with candor without fear of retaliation.

Servant Leadership: A leadership philosophy in which the goal of the leader is to serve others. Servant leaders believe in a structure in which the people/employees are at the top, and the leader is at the bottom.

Social Capital: Assigning value to the networks of relationships built among different people in a particular work or social setting. Social capital typically exists when there is a shared sense of identity and understanding. Strong social capital often includes shared values, norms, and trust.

Total Productive Maintenance (TPM): A process that uses production sheets, maintenance logs, modeling programs, and personnel assessments to assess the needs of company equipment and employees. This helps to keep machines and equipment running efficiently and at full capacity. It is an effective way to catch errors in a process before they become detrimental.

Values: The important things in one's life that dictate how one lives and works. They serve as a guide for setting priorities and working through difficult times. In an innovative seeing, values are demonstrated through the constructive action of those who are in leadership positions.

Value Stream Mapping (VSM): A visual tool used to understand and communicate the steps to change material or information for the purpose of creating a product or service that customers are willing to pay for, including the flow of value-adding activities that serve customer needs and the supply chain from raw material consumption.

Vision: Sometimes crafted in the form of a statement, this word refers to a future end state manifested by someone with the ability to plan for the future.

Index

Page numbers in *italics* refer figures

E

F

G

H

I

V

W

Printed in the United States
by Baker & Taylor Publisher Services